D0872297

OSCILLOSCOPE
HANDBOOK

OSCILLOSCOPE HANDBOOK

Clyde N. Herrick
San Jose City College

RESTON PUBLISHING COMPANY, INC.
Reston, Virginia
A Prentice-Hall Company

Library of Congress Cataloging in Publication Data

Herrick, Clyde N
  Oscilloscope handbook.

  1. Cathode ray oscilloscope—Handbooks, manuals, etc.
2. Electronic measurements—Handbooks, manuals, etc.
I. Title.
TK7878.7.H46   621.3815'48   73-21506
ISBN 0-87909-597-0

© 1974 by
Reston Publishing Company, Inc.
A Prentice-Hall Company
Reston, Virginia 22090

10 9 8

Printed in the United States of America.

CONTENTS

PREFACE

With the rapid advance of oscilloscope technology in recent years, a need has arisen for a relevant state-of-the-art handbook. Oscilloscopes have become more sophisticated and versatile. For example, triggered time bases with calibrated sweeps are rapidly replacing conventional free-running time bases. Vertical-input systems are often calibrated for rapid measurement of peak-to-peak voltage values. Frequency response from DC to 30 MHz, or higher, is common even in moderate-performance laboratory-type scopes. Dual-trace scopes have made entry in television service shops. Scopes with double-ended (push-pull) input facilities are available for making specialized tests. Various other special-purpose scopes have been developed.

It is the purpose of this oscilloscope handbook to present a broad coverage of all basic oscilloscope applications. Thus, fundamental laboratory, industrial, high-fidelity, multiplex, radio, and television applications are explained and illustrated. Basic measurements, signal-tracing procedures, and waveform analysis are discussed. Troubleshooting with the oscilloscope is explained along with basic oscilloscope theory. The focus of attention is on the beginning student and his problems. Students in advanced engineering courses will have acquired a background that has prepared them for assimilation of high academic-level texts. Mathematical treatment has been minimized in this handbook, to facilitate learning by students who may have a limited mathematical background.

It is assumed that the reader either has completed courses in electricity, electronics, radio communication, and television, or has attained a practical background in these areas. A student who is taking a concurrent course in black-and-white or color television will be able to assimilate the associated chapters in this book by looking up various technical terms and topics with which he may be unfamiliar. The burden on the student will be considerably lightened, however, if he has completed courses in both black-and-white and color television before starting his study of the oscilloscope.

Acknowledgment is made to those who have preceded the author by their development of other books on oscilloscope technology, and to the faculty of San Jose City College, who have made many helpful suggestions and criticisms. This book can be properly described as a team effort, although the individual members would choose to minimize the measure of their own contributions. It is appropriate that this handbook be dedicated as a teaching tool to the instructors and students of our junior colleges and technical schools.

Clyde N. Herrick

1

OSCILLOSCOPE
FUNDAMENTALS

1.1 INTRODUCTION

An oscilloscope is an electronic instrument consisting of a cathode-ray tube (CRT) and various associated circuit sections. As an illustration, Fig. 1-1 depicts a block diagram for a basic oscilloscope. Many authorities regard the oscilloscope as the most versatile of all electronic instruments. The oscilloscope is commonly used to automatically plot a particular voltage variation vs. time. Such a screen display is called a *waveform*. Current variation vs. time can also be displayed. Or a voltage variation vs. current variation can be automatically plotted. Phase relations of voltages and/or currents can be displayed. The output-voltage variation of a circuit vs. frequency can also be automatically plotted; this display is called a frequency-response curve. Oscilloscopes are often calibrated to indicate values of AC and DC voltages and currents. Many oscilloscopes are also calibrated to indicate periods of time.

A cathode-ray tube consists of an electron gun for supplying a concentrated beam of electrons, a pair of deflection plates for changing the

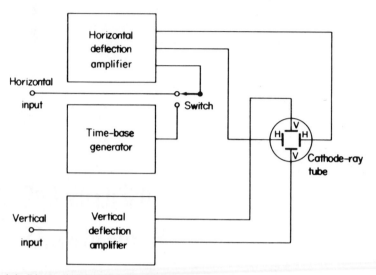

Fig. 1-1. Block diagram for a basic oscilloscope.

direction of the electron beam, and a screen coated with a substance that glows when struck by the electron beam. Some oscilloscopes are comparatively simple in construction, and have a minimum of internal circuit sections. For example, Fig. 1-2 shows the appearance of a comparatively simple oscilloscope. A transparent ruled screen, called a *graticule,* is generally mounted in front of the fluorescent screen. The CRT is con-

Fig. 1-2. A comparatively simple oscilloscope. (*Courtesy of* Heath Co.)

structed as shown in Fig. 1-3. Electrons can move from the cathode to the fluorescent screen because the tube has a high vacuum.

An electric current is passed through the heater of the electron gun in order to raise the temperature of the cathode sufficiently to emit electrons. The control electrode (grid no. 1 in Fig. 1-3) serves two important functions. First, this grid has a small orifice that provides an approximate "point" source of electrons from the electron gun. Second, the intensity of the electron beam is controlled by the bias voltage between this grid and the cathode. Evidently, if we apply a negative voltage to grid no. 1, more or less of the emitted electrons will be repelled back to the cathode, and will not emerge as an electron beam. On the other hand, if no bias voltage is applied to grid no. 1, the emitted electrons will be free to pass through the orifice and to proceed as an intense electron beam toward anode no. 1.

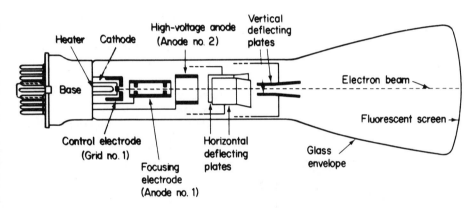

Fig. 1-3. Construction of a conventional cathode-ray tube.

There is no advantage in biasing grid no. 1 positive, because the grid would then collect the emitted electrons, and only a weak or perhaps no electron beam would proceed toward anode no. 1. It is evident that as grid no. 1 swings more or less negative, the intensity of the electron beam decreases and increases accordingly. The beam-current flow is comparatively small in most CRT's, and is measured in microamperes. A large beam current is not required, because a few microamperes produce ample light output from the fluorescent screen. An abnormally heavy beam current will "burn" the screen, requiring replacement of the CRT. It is good practice to avoid operation of the screen at unnecessarily high intensity.

With reference to Fig. 1-3, anode no. 1 is operated at a fixed positive voltage above the cathode potential. In turn, electrons that pass through the control grid are attracted (accelerated) toward the first anode. Like the control grid, the first anode is provided with small orifices that constrain the electron beam and help to keep its diameter small. As would be expected, diverging electrons in the beam are collected by the anode, with the result that there is a small anode no. 1 current flow. This is wasted current; however, it is quite small in well-designed CRT's.

We perceive in Fig. 1-3 that the electron beam which emerges from anode no. 1 consists of electrons that are traveling in practically parallel straight lines. The electron beam proceeds from anode no. 1 toward anode no. 2 because the second anode operates at a higher fixed positive voltage than the first anode. In effect, the electrostatic fields established by the DC voltages on the first and second anodes form an *electron lens* that can be compared with an optical lens. A visualization of an electron lens is seen in Fig. 1-4. Electrons entering the curved electric flux lines (field) between the anodes are subjected to inward-directed forces that act to focus the electron beam to a point. In turn, the electron paths converge to a point on the fluorescent screen in much the same way that a glass lens brings parallel rays of light to a point focus.

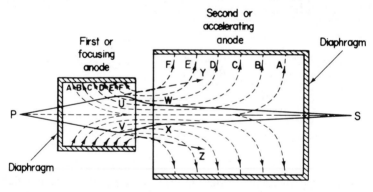

Fig. 1-4. Visualization of an electron lens.

When operating an oscilloscope, we bring the electron beam to a sharp focus by adjustment of the DC voltages on the anodes. Generally, the DC voltage on anode no. 2 is maintained at a fixed value, and the DC voltage on anode no. 1 is adjusted for optimum focus. Hence, the first anode is commonly called the *focusing electrode*. In the electrostatic type of CRT, such as depicted in Fig. 1-3, beam deflection is effected by means of two pairs of *deflecting plates*. The electron beam proceeds from the second anode through the deflecting plates and finally strikes the fluorescent screen, because the deflecting plates and screen are operated at a higher positive DC voltage than the second anode.

Note that the two pairs of deflecting plates are mounted at right angles to each other in the CRT, as shown in Fig. 1-5. Electrostatic deflecting fields are produced by application of positive and negative voltages between the two deflecting plates in each pair. In Fig. 1-5, these deflecting voltages are represented as batteries. We observe that the electron beam will be attracted toward the positive plate and repelled by the negative plate. In turn, if AC signal voltages are applied to the deflecting plates, a screen pattern (waveform) is traced by the electron beam.

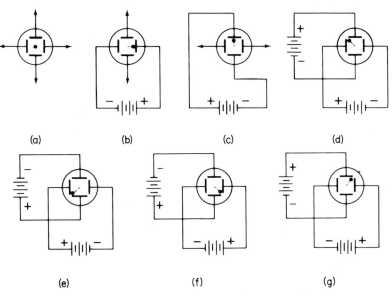

Fig. 1-5. Deflection of electron beam in CRT. (a) Both deflecting plates at zero voltage. (b) Positive voltage on right deflecting plate. (c) Positive voltage on upper deflecting plate. (d–g) Equal positive voltages on adjacent deflecting plates.

1.2 DC VOLTAGE SUPPLY FOR A CATHODE-RAY TUBE

Direct-current operating voltages for the CRT in an oscilloscope are provided by a suitable power supply. Figure 1-6 depicts a basic configuration. This is the most fundamental oscilloscope arrangement that can be used for waveform displays. Note that a 3AP1 CRT has a screen that is three

5

inches in diameter. The heater in the CRT is powered by a 2.5-volt winding on the transformer. Other electrode voltages are obtained from a high-
voltage winding with a half-wave rectifier diode and a filter capacitor.
This high-voltage winding provides 1120 volts rms, which has a peak
value of 1600 volts. Filter capacitor C charges up to 1600 volts at no
load, and operates at 1500 volts on full load.

Note that the high-voltage circuit is dangerous, and can cause injury or death if accidentally contacted. An oscilloscope should not be
operated with its case removed, except by an experienced technician. A
simple high-voltage filter is employed in Fig. 1-6 because the current
demand is very small. The power supply provides a negative output
voltage. Note that the deflecting plates and screen of the CRT operate
at zero volts with respect to ground, while the cathode operates at about
−1450 volts. This is done as a safety precaution. In other words, the
danger of shock to the operator is minimized by operating the screen at
ground potential. Insofar as electron acceleration is concerned, it makes
no difference whether the cathode is operated at ground potential and
the screen at +1450 volts, or whether the screen is operated at ground
potential and the cathode at −1450 volts. When an electron falls through
1000 volts, it is accelerated to approximately 10,000 miles per hour.

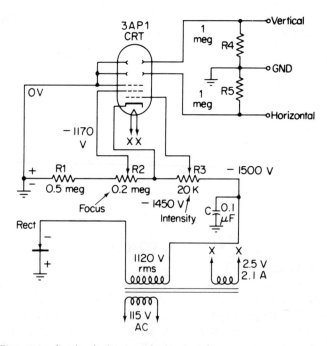

Fig. 1-6. Power-supply circuit for a cathode-ray tube.

Potentiometer R3 in Fig. 1-6 is adjusted to obtain the desired pattern brightness, or intensity. This adjustment varies the grid-cathode bias voltage in the CRT. Potentiometer R2 is adjusted to obtain maximum sharpness (focus) of the pattern. It adjusts the DC voltage on the first anode. If the intensity control is adjusted to apply about 75 volts bias to the control grid, the electron beam is "cut off," and the screen becomes dark. Observe that one of the vertical-deflecting plates and one of the horizontal-deflecting plates are connected to ground. This arrangement provides for application of single-ended deflection voltages to the deflecting plates. That is, the deflection voltages or signals that we apply to the deflecting plates will be provided by a pair of leads, one of which is grounded, and the other of which is "hot."

Resistors R4 and R5 in Fig. 1-6 have comparatively high values (1 megohm). Their function is merely to reference the right-hand deflecting plates to ground potential. Beam deflection is obtained by applying input voltages to each of the deflecting plates indicated by "vertical" and "horizontal" terminals. If DC deflection voltages are applied, we shall obtain stationary beam displacements, as depicted in Fig. 1-5. On the other hand, if AC deflection voltages are applied, we obtain various screen patterns, as explained in the following section. In either case, approximately 110 volts are required to deflect the electron beam 1 inch on the screen. Or about 330 volts are required to obtain full-screen deflection in a 3AP1 CRT.

1.3 SCREEN PATTERNS OBTAINED WITH AC DEFLECTION VOLTAGES

With no deflection voltages applied to the vertical- and horizontal-input terminals in Fig. 1-6, the spot rests at the center of the screen, as depicted in Fig. 1-7. Next, if we apply an AC signal voltage with an ampli-

Fig. 1-7. Spot rests at center screen in absence of vertical and horizontal input signals.

7

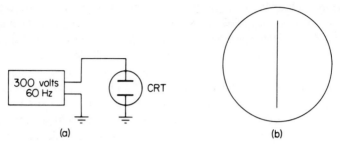

Fig. 1-8. (a) Test setup. (b) Display of vertical straight line.

tude of approximately 300 volts between the vertical-input terminal and ground, a vertical line is displayed on the screen, as shown in Fig. 1-8. If the frequency of the AC signal voltage is 60 Hz, the spot moves up and down 60 times a second. However, no flicker is observed and the line is displayed steadily because of the persistence of vision inherent in the human eye. Next, if the signal voltage is applied between the horizontal-input terminal and ground, a horizontal line will be displayed on the screen.

It is important to know the voltage relations in a sine wave, as indicated in Fig. 1-9. Root-mean square (rms), positive-peak, negative-peak, and peak-to-peak values are shown. Most AC voltmeters read the rms values of sine waves. In turn, when the waveform depicted in Fig. 1-9 is applied to a conventional AC voltmeter, the meter will indicate 1 volt. Note that some transistor voltmeters provide both rms and peak indicating

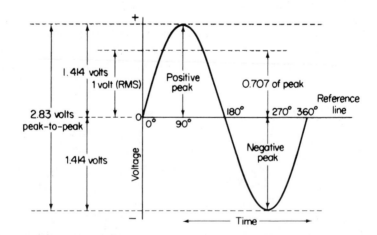

Fig. 1-9. Voltage relations in a sine wave.

scales. Accordingly, an AC voltmeter of this type will indicate 1.414 volts on its peak-voltage scale for the waveform of Fig. 1-9. Again, various transistor voltmeters provide peak-to-peak indication. An AC voltmeter of this type will indicate 2.83 volts on its peak-to-peak voltage scale for the waveform of Fig. 1-9. The rms value of a sine wave is sometimes called its effective value. This means that the effective (rms) value produces the same amount of power as a DC voltage of the same value.

Insofar as deflection of the electron beam in a CRT is concerned, a peak-to-peak voltage is equivalent to a DC voltage of the same value. As an illustration, if we substitute a 300-volt DC source for a 300-volt peak-to-peak (p-p) AC source, the electron beam in a CRT will be deflected by the same amount in either case. Of course, a DC source does not provide sustained up-and-down motion of the beam, unless the DC voltage is switched off and on repeatedly. Note that, as shown in Fig. 1-10, a +150-volt potential produces half-screen deflection upward, and a −150-volt potential produces half-screen deflection downward. A summary of the voltage relations for a sine waveform follows:

Peak volts = 1.414 × rms volts
RMS volts = 0.707 × peak volts
Peak-to-peak volts = 2.83 × rms volts
Peak-to-peak volts = 2 × peak volts
(Positive and negative peak voltages are equal in a sine wave.)

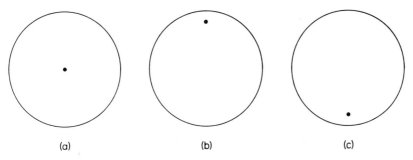

(a)   (b)   (c)

Fig. 1-10. Response of CRT beam to DC voltages. (a) Zero voltage applied. (b) Upper deflecting plate 150 volts positive; lower plate grounded. (c) Upper deflecting plate 150 volts negative; lower plate grounded.

It follows that an oscilloscope can be applied as a voltmeter. As an illustration, if the spot deflects 1 inch on the screen for an applied voltage of 110 volts DC, or 110 volts p-p AC, the *deflection sensitivity* of the oscilloscope is 110 volts p-p per inch. In turn, if an unknown value of applied voltage deflects the spot 2 inches, its indicated value is 220 volts p-p. Or, if an unknown voltage deflects the spot ½ inch, its indicated value is 55 volts p-p. This topic is explained in greater detail subsequently under the heading of calibration procedures.

Next, it is instructive to consider frequency-measurement procedures by means of *Lissajous figures*. With reference to Fig. 1-11, a sine-wave voltage source is applied to both the vertical- and horizontal-deflecting plates of the CRT. In turn, a diagonal straight line is displayed on the screen. Note that the vertical- and horizontal-deflection voltages are necessarily in phase with each other (go through zero at the same instant). This is the requirement for display of a straight diagonal line. Note also that the vertical- and horizontal-deflection voltages are necessarily equal in amplitude. This is the requirement for display of a straight line at a

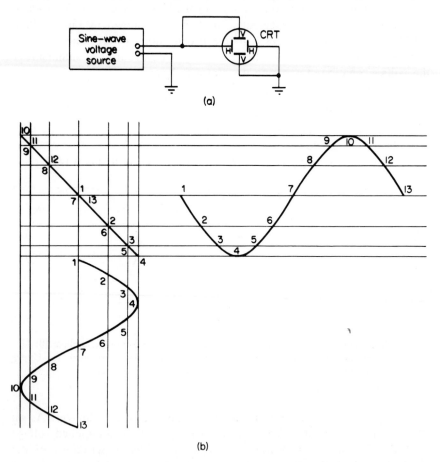

Fig. 1-11. (a) Sine-wave voltage applied to both pairs of deflecting plates. (b) Screen pattern is a diagonal straight line.

10

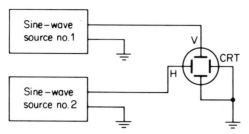

Fig. 1-12. Individual sine-wave voltages applied to vertical- and horizontal-deflecting plates.

45-deg angle. This same display will be produced by the test setup depicted in Fig. 1-12, provided that the two sine-wave sources provide outputs that are in phase with each other and of equal amplitude.

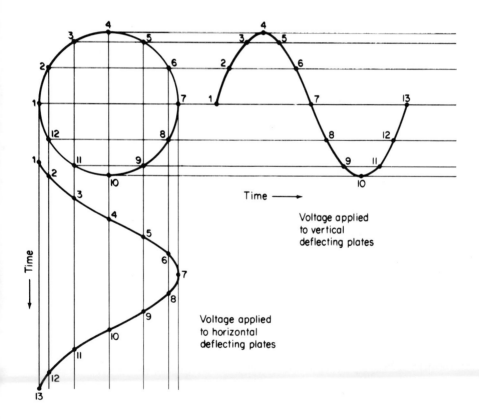

Fig. 1-13. Development of a circular pattern by two sine waves with the same frequency and amplitude, but 90 deg different in phase.

Now, if the sine-wave sources in Fig. 1-12 have the same amplitude and the same frequency, but are 90 deg different in phase, a circular Lissajous figure is displayed on the CRT screen, as depicted in Fig. 1-13.

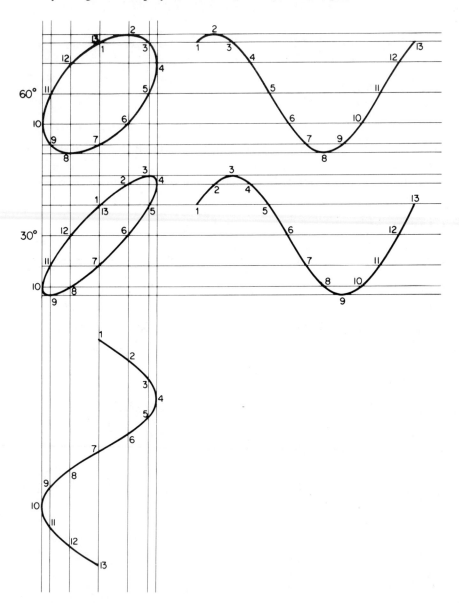

Fig. 1-14. Elliptical Lissajous figures produced by two sine waves with the same frequency and amplitude, but with 30-deg and 60-deg phase differences.

As would be anticipated, phase differences in the range from 0 to 90 deg produce elliptical Lissajous figures, as exemplified in Fig. 1-14. Any phase angle can be measured as shown in Fig. 1-15. That is, the ellipse is carefully centered on the screen, and the intervals M and N are measured. Then, the phase angle between the vertical- and horizontal-deflecting voltages is equal to arcsin M/N. Figure 1-16 shows the progression of patterns in this situation for a range of 360 deg in 45-deg steps. Note that the 45-deg ellipse leans to the right, whereas the 135-deg ellipse leans to the left.

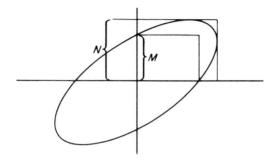

Fig. 1-15. Phase-angle difference of the deflection voltages is equal to arcsin M/N.

<table>
<tr><td>0°
360°</td><td>45°
315°</td><td>90°
270°</td><td>135°
225°</td><td>180°</td></tr>
</table>

Fig. 1-16. Lissajous figures in the range from 0 deg to 360 deg.

We shall now consider Lissajous patterns produced by vertical- and horizontal-deflection voltages that have the same frequency, a phase difference of 90 deg, and unequal amplitudes. With reference to Fig. 1-17, the ellipses that are displayed have their axes in vertical and horizontal directions. This is the basic distinction between ellipses produced by unequal voltage amplitudes, and ellipses produced by voltages with a phase difference. In other words, the ellipse depicted in Fig. 1-15 has its axes inclined at 45 deg to the vertical and horizontal directions—this is always true for any phase difference, provided that the voltages have equal amplitudes. Note also that even if the voltages have unequal amplitudes, the phase-angle measurement is always given by arcsin M/N.

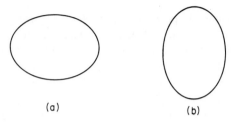

(a)   (b)

Fig. 1-17. Ellipses produced by unequal signal voltages having a phase difference of 90 deg. (a) Horizontal voltage greater than vertical voltage. (b) Vertical voltage greater than horizontal voltage.

Next, with reference to Fig. 1-18, we shall consider Lissajous patterns produced by sine-wave voltages that have equal amplitudes, but which differ in frequency by a 2-to-1 ratio. If the input voltages are in phase, the pattern appears as shown in (a). Or, if the input voltages are 90 deg out of phase, the loops are closed as seen in (c). Other phase relations are exemplified in (b) and (d). Note that if the frequency ratio of the input voltages is exactly 2 to 1, the resulting Lissajous pattern will not change. On the other hand, if the frequency ratio is slightly different from 2 to 1, the pattern will change progressively through the shapes depicted in Fig. 1-18. A larger departure from a 2-to-1 frequency ratio will cause the pattern to change more rapidly through its intermediate forms.

There is a general rule that can be applied to any of the four Lissajous patterns in Fig. 1-18, to determine that the indicated frequency ratio is 2 to 1. This rule states that the frequency ratio is given by the number of tangencies to vertical and horizontal boundaries of the pattern. For ex-ample, note that there are two horizontal tangencies in (a), (b), and (d). Again, there is one vertical tangency in (a), (b), and (d). The pattern in (c) is a special case, in that the two halves are superimposed. Note that in the example of Fig. 1-18, the vertical-deflection frequency is double the horizontal-deflection frequency. In case the vertical and horizontal inputs are interchanged, the resulting Lissajous pattern will be turned through 90 deg on the screen.

With reference to Fig. 1-19, we observe a Lissajous pattern from input voltages having equal amplitudes, a frequency ratio of 3 to 2, and a phase difference of zero. The frequency ratio is indicated by the fact that the pattern is tangent to its horizontal boundary at two points, and is tangent to its vertical boundary at three points. Other basic Lissajous patterns are exemplified in Fig. 1-20. Such patterns are commonly used to calibrate audio oscillators against a known frequency source, such as the 60-Hz power line. It is usually impractical to make calibration checks at frequency ratios greater than 10 to 1, because of pattern instability.

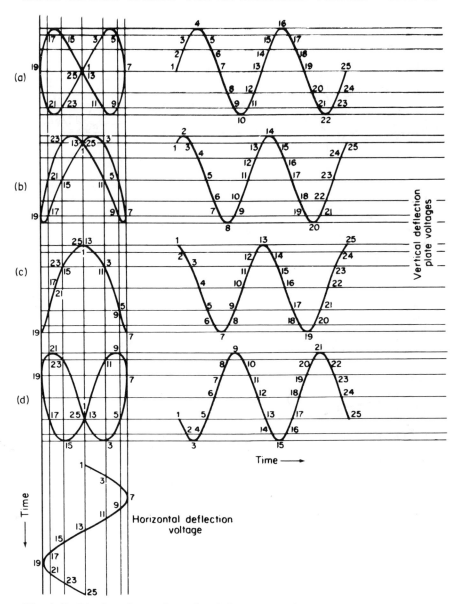

Fig. 1-18. Lissajous figures for a 2-to-1 frequency ratio.

is tangent to its vertical boundary at three points. Other basic Lissajous figures will provide easily interpreted patterns for high-frequency waveforms. An example is shown in Fig. 1-21 for typical *percentage-modula-*

15

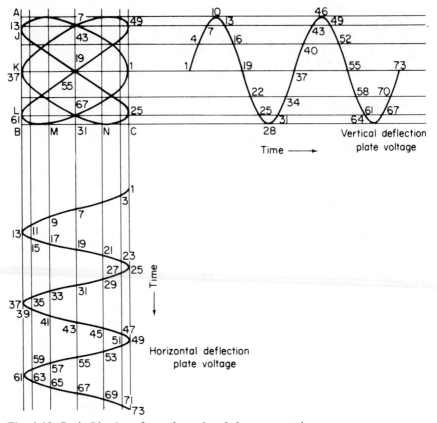

Fig. 1-19. Basic Lissajous figure for a 3-to-2 frequency ratio.

tion patterns. These are called trapezoidal patterns; percentage modulation is indicated by the pattern outline. If there is no modulation of the carrier (0 per cent modulation), a vertical line is displayed. Trapezoidal patterns are displayed in the range from 0 to 100 per cent modulation, and at 100 per cent the trapezoid becomes a triangle. At greater than 100 per cent modulation (overmodulation), the triangle is followed by a horizontal line.

Figure 1-22 shows how this type of Lissajous figure is developed. The RF carrier is amplitude-modulated by a single audio frequency. The RF carrier voltage is applied to the vertical-deflecting plates, and the audio voltage is applied to the horizontal-deflecting plates in the CRT. This type of test is practical at very high carrier frequencies, because a CRT is responsive to frequencies of many megahertz. Figure 1-23 shows a step-by-step development of a trapezoidal pattern. Note that only sine-wave

16

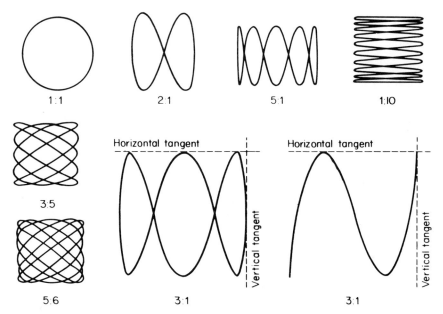

Fig. 1-20. Lissajous figures with various frequency ratios.

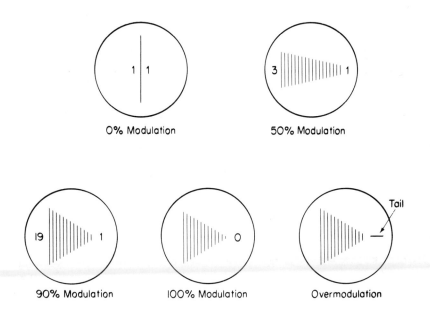

Fig. 1-21. Lissajous patterns indicating various percentages of amplitude modulation.

17

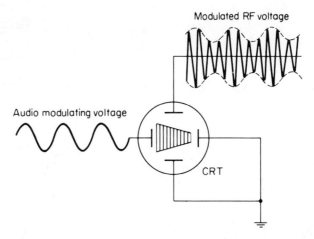

Fig. 1-22. Basic arrangement of the percentage-modulation test setup.

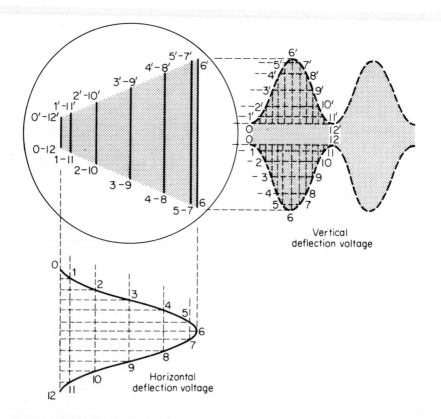

Fig. 1-23. Step-by-step development of a trapezoidal pattern.

modulation is suitable; if the sine waveform happens to be distorted, the trapezoidal pattern will also be distorted.

A typical test setup for display of trapezoidal patterns is shown in Fig. 1-24. The deflection voltages are obtained by coupling the CRT to the RF amplifier tank and to the modulator plate in the transmitter. An RF pickup loop with several turns is employed; it is spaced from the tank coil at a distance that provides adequate vertical deflection on the CRT screen. Note that a dangerous arc may be drawn if the pickup loop is spaced too closely to the tank. More turns should be utilized on the loop, if necessary. Capacitive coupling is used to the modulator plate, with a potentiometer for adjustment of horizontal deflection. Note that C1 must have a voltage rating at least double the plate voltage on the modulator tube. Otherwise, the capacitor is likely to break down and short-circuit the plate-voltage supply into the test setup.

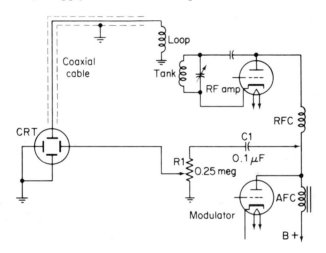

Fig. 1-24. Test setup for display of trapezoidal patterns.

Consideration of pattern development over a complete cycle in Fig. 1-23 will show that the trapezoid is first traced from left to right by the electron beam, and is then repeated from right to left. If the two deflection voltages are correctly phased, the left-hand and right-hand trapezoids occupy the same area on the screen (lay over correctly). On the other hand, if the audio modulating voltage is more or less out of phase with the envelope of the modulated RF voltage, an offset type of double-image display results, as exemplified in Fig. 1-25. Correct phase adjustment is obtained by turning the loop in Fig. 1-24 more or less at an angle with the tank, as required. Note that pattern distortion is sometimes

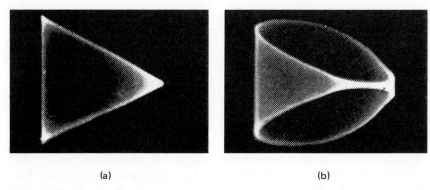

(a)   (b)

Fig. 1-25. (a) Correct display of a trapezoidal pattern. (b) Poor layover owing to phase error.

caused by substantial harmonic distortion from the tank, or by an improperly filtered transmitter power supply.

We shall find that Lissajous patterns are used in various other electronic test procedures. In most cases, these applications require an oscilloscope with a suitable vertical amplifier, horizontal amplifier, or both. The reason for this is that the test signals available in many types of electronic equipment do not have sufficient amplitude (voltage) to deflect the beam in a CRT directly. Details of oscilloscope amplifiers are explained in the following chapters. Figure 1-26 shows the locations of the vertical and horizontal amplifier control zones.

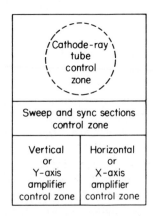

Fig. 1-26. The vertical and horizontal amplifier control zones are usually at the lower left- and right-hand areas of the panel.

1.4 INTRODUCTION TO SAWTOOTH WAVEFORMS

In Fig. 1-1, a time-base generator was depicted. This is basically a sawtooth waveform generator, which is usually built into the oscilloscope. When a sawtooth voltage is applied to the horizontal-deflecting plates and a sine-wave voltage is applied to the vertical-deflecting plates of a CRT, as depicted in Fig. 1-27, a sine-wave pattern will be displayed on the oscilloscope screen. The reason for this pattern development is

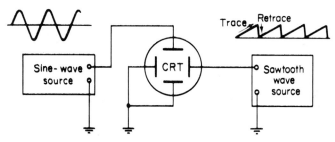

Fig. 1-27. Arrangement for display of a sine-wave pattern on a CRT screen.

that a sawtooth waveform has a voltage that increases in direct proportion to time. This is the origin of the term "time base." When a sine waveform is displayed on the CRT screen, the oscilloscope operating controls have characteristic effects on the pattern display, as exemplified in Fig. 1-28.

At this point in our study of the oscilloscope, we can understand some of these control functions. On the other hand, we can form only a very general idea of other control functions. A few of the functions necessarily seem puzzling and meaningless at this time. Therefore, we must proceed to build upon what we know, and to increase our understanding of the oscilloscope from both theoretical and practical considerations. It is particularly helpful at this point to observe carefully the following demonstrations that will be made by your instructor. Or, if an instructor is not available for this purpose, you may conduct the experiments by yourself.

Demonstration

SUBJECT

The cathode-ray oscilloscope.

21

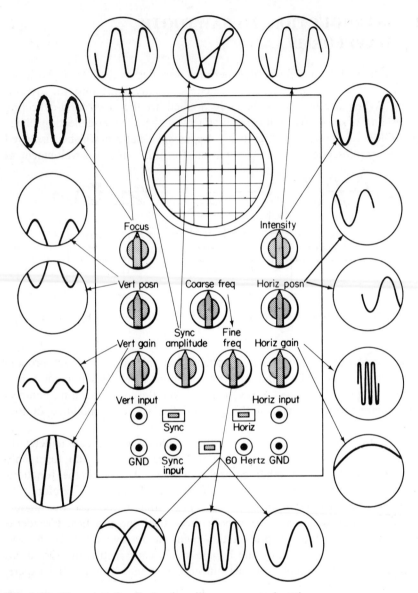

Fig. 1-28. Characteristic effects of oscilloscope control settings.

OBJECTIVE

To familiarize the student with the names and basic operation of oscilloscope controls, and to demonstrate some fundamental applications of the oscilloscope.

Material Required

1. Simple oscilloscope.
2. Audio oscillator.
3. Test leads.

Points To Be Stressed

A spot should not be allowed to remain on the CRT screen with the intensity control turned to an excessively high position. Otherwise, the intense electron beam will burn a hole in the coating of the screen, causing a permanent brown spot to appear. A general operating rule is to keep the setting of the intensity control no higher than is actually required.

Procedure

Operation of Controls

1. Explain the names and point out the locations of the various operating controls.
2. Plug the cord into an outlet, and explain the necessity for a warm-up interval. Explain why the intensity control is turned fully counterclockwise during the warm-up period.
3. Turn the intensity control slowly clockwise, instructing the students to watch the horizontal line that is produced on the CRT screen by action of the sweep oscillator. (If a horizontal line does not appear, the sweep oscillator may not be connected to the horizontal-amplifier section.) In such a case, explain that we must check the horizontal function switch, and set it to its internal-sweep (or equivalent) position. A horizontal line will now be displayed on the CRT screen. Explain to the students that when the switch is set to its direct position, the sweep oscillator is disconnected from the horizontal-amplifier section—the input of the horizontal amplifier is now connected to the binding posts marked "horizontal input" on the front panel of the oscilloscope.
4. Vary the intensity and focus controls, and point out how the brilliance and sharpness of the displayed line are thereby affected. Show how these controls may tend to interact, and how to set them to the positions that provide the sharpest displayed line with adequate brilliance.
5. Connect a pair of test leads to the vertical-input terminals of the oscilloscope, and to the output terminals of an audio oscillator. Explain to the students that the ground lead from the audio

oscillator must be connected to the ground terminal of the oscilloscope.

6. Vary the vertical-gain control of the oscilloscope and point out that the screen pattern becomes larger or smaller in vertical amplitude, but does not change its size in the horizontal direction.

7. Vary the output level of the audio oscillator, to show that the same pattern variation occurs as in the preceding step.

Linear Sweep Patterns

1. Show the students how the audio oscillator is set to a frequency of 60 Hz. Point out how to set the oscilloscope controls as follows:

SYNC	Internal
FREQ	60 Hz
HORIZ GAIN	To provide a convenient length of horizontal trace
SYNC (LOCKING)	Zero
VERT GAIN	To provide a vertical display amplitude of one inch

2. Demonstrate how the frequency control is adjusted to display a single sine-wave cycle on the CRT screen.

3. Show how a fine adjustment of the frequency control is made to hold the pattern nearly stationary. Then, point out how the locking control is advanced to hold the pattern completely stationary.

4. Turn the frequency dial of the audio oscillator in the required steps to display 2, 3, and then 4 sine waves on the CRT screen.

5. Return the frequency dial of the audio oscillator to 60 Hz, and and then explain to the students that the signal frequency is to be further increased to a value that will display two lines resembling a pair of ice tongs on the CRT screen.

6. Have the students calculate the frequency of the audio oscillator when the sine wave. is displayed, and when each of the patterns in steps 4 and 5 is observed.

7. Disconnect the ground lead from the oscilloscope, and have the students sketch the resulting pattern. Ask them to explain, if possible, the result of an open ground connection in the test setup.

Lissajous Figures

1. Explain to the students that the ground lead will now be connected

to the oscilloscope, and that the output ("hot") lead of the audio oscillator will be connected to both the vertical-input terminal and the horizontal-input terminal of the oscilloscope.

2. Set the horizontal-selector control so that the horizontal-input terminal of the oscilloscope is connected to the input of the horizontal amplifier.

3. Have the students sketch the pattern that is displayed. Ask them to explain how the diagonal line is developed and its significance.

4. Apply a 60-Hz sine-wave voltage to the horizontal-input terminals of the oscilloscope.

5. With the audio oscillator connected to the vertical-input terminals of the oscilloscope, vary the oscillator frequency slowly in the vicinity of 60 Hz and point out the changing screen pattern to the students. Ask them to explain how the changing pattern is being formed.

6. Set the audio oscillator to 60, 120, 180, 30, and 20 Hz. Have the students sketch each of the resulting screen patterns.

7. Have the students discuss which method they consider to be more accurate for determination of the frequency of an audio signal: the Lissajous-figure method or the linear-sweep method. Ask them to explain why they arrived at the particular conclusion.

Synchronization

1. Set the audio oscillator at 100 Hz and display a single sine-wave cycle on the CRT screen. Ask the students at what frequency the oscilloscope sweep oscillator is now operating.

2 Turn the gain/sync control to approximately mid-position and turn the function control through its complete range. Have the students observe the point at which the displayed pattern becomes stationary, and have them attempt to explain this operating condition.

3. Reconnect the equipment as in step 1 under the Lissajous-figure demonstration. Turn the function control of the oscilloscope through its complete range. Have the students observe the points at which the displayed pattern becomes stationary. Ask the students to suggest the operating conditions that are involved.

Conclusions

1. From the demonstrations that have been made, and from the material that the students have learned in reading the first chapter of

25

the text, ask them to draw a block diagram of a simple oscillo-scope.

2. Have the students explain how a vertical sine wave and a horizon-tal sawtooth wave operate to display two sine-wave cycles on the CRT screen.

3. Ask the students to sketch the development of a circle pattern on the CRT screen, with sine-wave vertical- and horizontal-input voltages.

4. Let the students suggest several applications for an oscilloscope.

5. Request a list of the oscilloscope controls, with the basic function of each.

2

HORIZONTAL-
DEFLECTION
SYSTEMS

2.1 HORIZONTAL-DEFLECTION SYSTEMS

As explained in the first chapter, a cathode-ray oscilloscope is an electronic measuring device that displays electrical relationships as waveforms. Most waveforms are displayed as a plot or graph of voltage variation in time. That is, the horizontal movement of the electron beam is proportional to time, and the vertical movement of the beam is proportional to voltage. The time reference (time base) is generally built into the oscilloscope. This time base is also called a sweep generator. It produces a sawtooth voltage for horizontal beam deflection, as depicted in Fig. 2-1. A sawtooth voltage waveform is required to produce a uniform horizontal motion of the spot (linear deflection).

Figure 2-2 shows the appearance of the CRT screen with the time base operating, but with no vertical-input signal applied; the illustration also shows how a sine-wave signal is displayed when the frequency of the saw-tooth waveform is equal to the frequency of the sine-wave vertical-input signal. If the time base has one-half the frequency of the sine wave, two

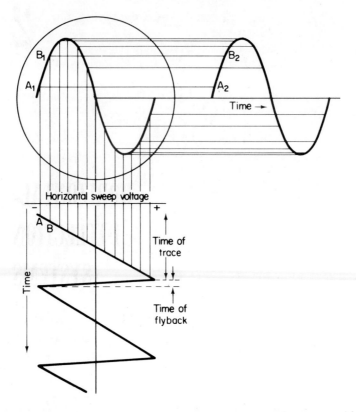

Fig. 2-1. Display of a sine waveform, using sawtooth horizontal deflection.

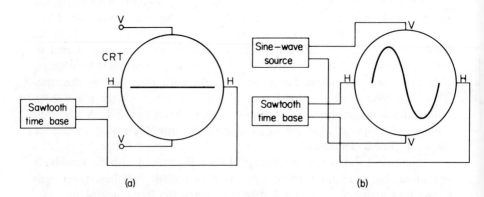

Fig. 2-2. (a) Time base operating, with no vertical-input signal. (b) Time base operating with sine-wave vertical-input signal.

cycles of the sine wave will be displayed. Or, if the time base has one-third the frequency of the sine wave, three cycles of the sine wave will be displayed, and so on. Again, if the frequency of the sine wave is double the frequency of the time base, two cycles of the sine wave will be displayed. Or, if the frequency of the sine wave is three times the frequency of the time base, three cycles of the sine wave will be displayed, and so on. However, if the time-base frequency is greater than the sine-wave frequency, pattern distortion results, as seen in Fig. 2-3.

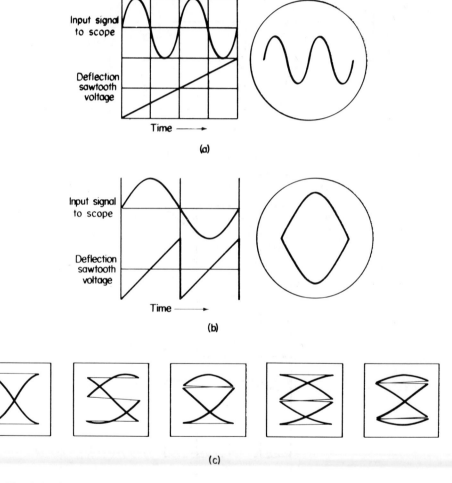

Fig. 2-3. Sine-wave displays at various time-base frequencies. (a) Time-base frequency is one-half of the sine-wave frequency. (b) Time-base frequency is double the sine-wave frequency. (c) Other displays caused by operation of the time base at too high frequencies.

29

2.2 BASIC TIME-BASE OSCILLATORS

All time-base oscillators are derived from the fundamental multivibrator configuration shown in Fig. 2-4. A multivibrator is essentially an amplifier

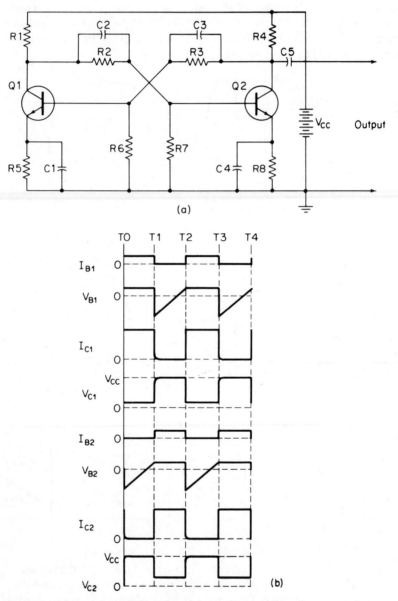

Fig. 2-4. Basic transistor multivibrator configuration. (a) Schematic. (b) Waveforms.

30

that provides its own input. That is, the output from transistor Q2 is coupled to the input of Q1. In turn, the transistors operate in the switching mode. One transistor is cut off while the other conducts; as the cutoff transistor goes into conduction, the conducting transistor cuts off. This sequence of circuit action is seen in the operating waveforms. Note that the collector waveforms are basically square waves, and that the base waveforms have a sawtooth component. This arrangement is called a free-running or astable multivibrator, because it oscillates by itself, without the necessity for any external control or trigger pulses.

The oscillating frequency of the multivibrator depicted in Fig. 2-4 is determined by the time constants of the base-coupling RC circuits R2C2 and R3C3. That is, the time that Q1 remains cut off during a cycle of operation depends on the time required for a charge on C3 to decay through R3. Similarly, the time that Q2 remains cut off depends on the time required for a charge on Q2 to decay through R2. This circuit action is apparent in waveforms V_{B1} and V_{B2}. If, for example, the value of C2 is 500 pf, and the value of R2 is 1 megohm, the time constant of the RC circuit is 0.0005 second. Figure 2-5 shows a universal time-constant chart for RC circuits. The time constant in seconds is equal to the product of resistance in ohms times the capacitance in farads.

Observe in Fig. 2-4 that the multivibrator will become a pulse generator if the time constant of R2C2 is quite short, and the time constant of R3C3 is quite long. This arrangement is called an unsymmetrical multivibrator. In an oscilloscope time base, one time constant is made as short as possible, and the other time constant is adjustable, in order to obtain a range of oscillating frequencies. As an illustration, Fig. 2-6 depicts the basic configuration for a source-coupled sawtooth oscillator. Q1 is coupled to Q2 via C2R2, and Q2 is coupled to Q1 via R. Since the time constant provided by the common source resistor R is very short, this arrangement operates as a pulse generator. A pulse is a short surge of voltage. The time between successive pulses is determined by the time constant of R2C2.

In order to obtain a sawtooth waveform output from the pulse generator in Fig. 2-6, capacitor C_o is shunted from the drain of Q2 to ground. Note that while Q2 is cut off, C_o charges from E_B through r_2, thereby forming the rising portion of the sawtooth (ramp). Next, when Q2 goes briefly into conduction, the charge on C_o is suddenly conducted to ground through Q2 and R. This discharge thus forms the rapid flyback portion of the sawtooth. It is apparent in Fig. 2-5 that a sawtooth waveform generated by an RC circuit cannot be perfectly linear (have a straight-line rise). In practice, only a small interval of curve A is employed in sawtooth generation (from 0 to 5 per cent, for example). In turn, the sawtooth is sufficiently linear for all practical purposes. This limits the ampli-

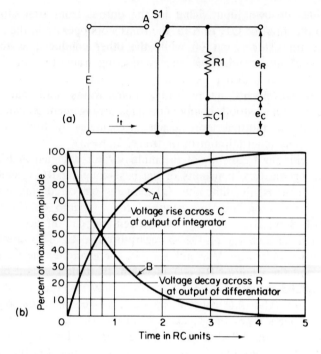

Fig. 2-5. Universal time-constant chart for RC circuits. (a) Basic circuit arrangement. (b) Chart.

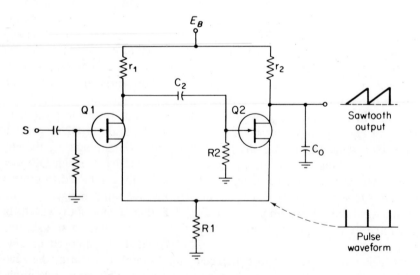

Fig. 2-6. Source-coupled sawtooth oscillator.

tude of the sawtooth, and an amplifier is utilized to step up its amplitude for full-screen deflection of the CRT beam.

We observe in Fig. 2-6 that the sawtooth frequency can be changed by varying the value of C2. To maintain reasonably uniform amplitude of the generated waveform, the value of C_o must be changed correspondingly. That is, when the value of C2 is increased, the value of C_o is also increased. If the value of C_o is not increased, a greater output amplitude results, but the sawtooth also becomes excessively nonlinear. When a nonlinear saw-tooth is used to deflect the CRT beam, the resulting pattern display becomes cramped at one end, as exemplified in Fig. 2-7. Note that improved linearity can be obtained by increasing the supply voltage E_B in Fig. 2-6. This restricts the operating range to a smaller interval of curve A in Fig. 2-5. However, a practical limit is imposed on the value of E_B by the rating of transistors Q1 and Q2 in Fig. 2-6.

Synchronization of a time base is required in order to lock a displayed waveform in place on the CRT screen. In the absence of synchronization,

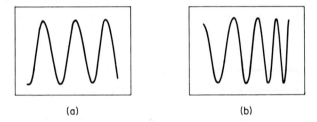

(a)   (b)

Fig. 2-7. Displays of a sine waveform. (a) Comparatively linear sawtooth deflection. (b) Nonlinear sawtooth deflection.

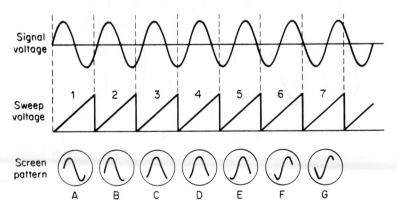

Fig. 2-8. Drift of a CRT screen pattern in the absence of synchronization. (*Courtesy of* U.S. Armed Forces).

33

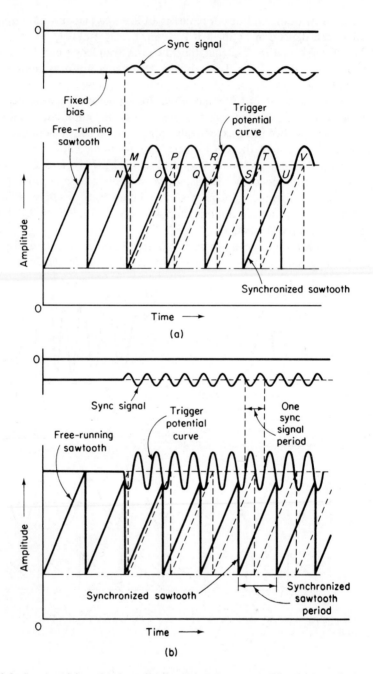

Fig. 2-9. Synchronizing circuit action in a time base. (a) For display of one cycle on CRT screen. (b) For display of two cycles.

the pattern will move to the right or to the left instead of remaining stationary. This condition is depicted in Fig. 2-8 for the case in which the time base operates at a slightly higher frequency than that of the vertical-input signal. Synchronization is obtained by coupling a sample of the vertical-input signal to the gate of Q1 in Fig. 2-6. In practice, the time base is adjusted to operate at a slightly lower frequency than that of the vertical-input signal. In turn, Q1 is triggered out of cutoff slightly earlier than in its free-running mode, as shown in Fig. 2-9. Thereby, the pattern is stabilized on the CRT screen, and is said to be locked in sync. Note that the synchronized sawtooth oscillator comes out of cutoff at instants N, O, Q, etc., instead of instants $M, P, R,$ etc.

It is necessary to apply sufficient sync-signal amplitude to the gate of Q1 (Fig. 2-6) to obtain reliable triggering action. As an illustration, if the sync-signal amplitude is too weak, the displayed waveform tends to jitter horizontally, to drift, or to lose sync lock entirely, as exemplified in Fig. 2-10. On the other hand, if the sync-signal amplitude is excessive, the waveform display will be distorted, as exemplified in Fig. 2-11. Therefore, an oscilloscope is provided with a sync-amplitude control to enable correct locking action.

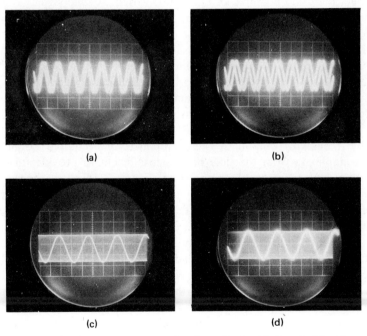

(a) (b)

(c) (d)

Fig. 2-10. Examples of insufficient sync-signal amplitude. (a) Marginal locking action. (b) Locking action almost lost. (c–d) Locking action completely lost.

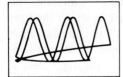 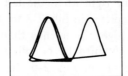 

Fig. 2-11. Examples of distorted displays due to excessive sync-signal amplitude. (*Courtesy of* U.S. Armed Forces)

2.3 COMPLETE TIME-BASE OSCILLATOR

A complete time-base oscillator is depicted in Fig. 2-12. It provides a free-running sawtooth output from 20 Hz to 100 kHz, and a blanking-pulse output for eliminating the retrace line in the screen pattern. The basic circuit is the same as shown in Fig. 2-6. However, capacitors C14 through C19 are provided for step (coarse) change of the oscillator frequency. Vernier (fine) control of frequency is provided by the ganged potentiometers R30A and R30B. As the time constant of the gate circuit is varied, the drain potential of Q2 is also varied. A sync signal is applied to the gate of Q1. Amplitude of the sync signal is adjustable by means of potentiometer R16. A choice of internal or external sync source is provided. Internal sync is taken from the vertical-deflection signal; external sync may be applied from any arbitrary source. Details are explained subsequently. Note that an external horizontal-deflection signal may be applied to the horizontal-input terminal when sawtooth deflection is not employed, as for display of Lissajous figures.

As seen in Fig. 2-13, the CRT is driven by a horizontal amplifier and by a blanking amplifier. These amplifiers are required because the output amplitude from the time base is insufficient to provide full-screen deflection, and adequate blanking action. Figure 2-14 shows how a sine-wave display is expanded horizontally as the gain of the horizontal amplifier is increased. Figure 2-15 illustrates the effect of retrace blanking in a displayed waveform. Note that Fig. 2-13 depicts direct coupling from the horizontal amplifier and the blanking amplifier to the CRT. In various oscilloscopes, AC coupling is utilized instead, as exemplified in Fig. 2-16. When AC coupling is utilized, the electron beam in the CRT cannot be deflected by a DC voltage source. Instead, the CRT input circuit has a lower-frequency cutoff limit, such as 20 Hz. On the other hand, when DC coupling is used, the CRT input circuit has no lower-frequency cutoff, and the beam can be deflected by a DC voltage source.

Note also in Fig. 2-16 that vertical- and horizontal-centering controls are shown. These controls serve to adjust the DC bias voltage on the

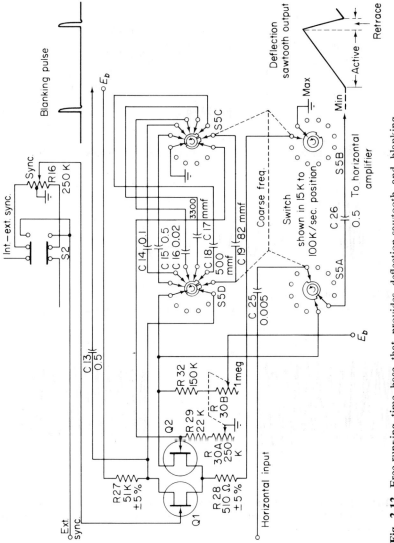

Fig. 2-12. Free-running time base that provides deflection-sawtooth and blanking-pulse outputs.

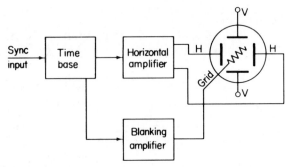

Fig. 2-13. CRT is driven by a horizontal amplifier and by a blanking amplifier.

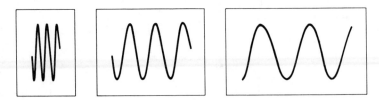

Fig. 2-14. A displayed waveform is expanded horizontally as the gain of the horizontal amplifier is increased. (*Courtesy of* U.S. Armed Forces)

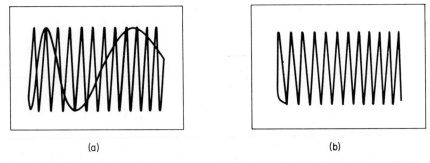

(a)   (b)

Fig. 2-15. Effect of blanking on retrace visibility. (a) No blanking pulse applied to CRT. (b) Retrace eliminated by blanking pulse. (*Courtesy of* U.S. Armed Forces)

CRT deflecting plates, so that the pattern can be shifted on the screen as desired. Figure 2-17 shows the basic action of the centering controls. The focus control is adjusted for the smallest and sharpest spot (or trace), as seen in Fig. 2-18. It is important to avoid turning the intensity control so high that the spot or trace becomes excessively bright and burns the screen. The lower-frequency limit of the CRT input circuitry in Fig. 2-16 is determined by the time constants of the RC coupling networks. Thus, C3

38

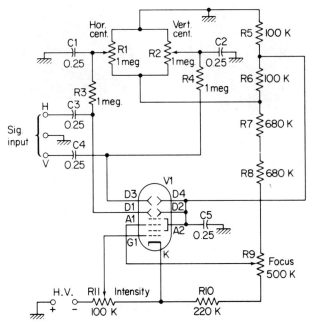

Fig. 2-16. Example of AC coupling to CRT deflecting plates.

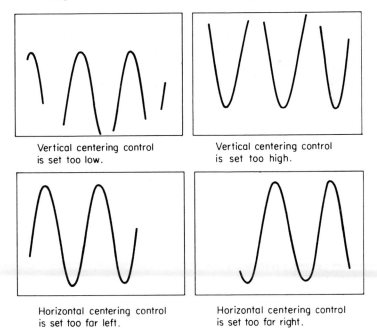

Vertical centering control
is set too low.

Vertical centering control
is set too high.

Horizontal centering control
is set too far left.

Horizontal centering control
is set too far right.

Fig. 2-17. Basic action of the centering controls.

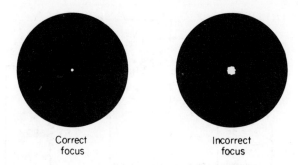

Correct
focus

Incorrect
focus

Fig. 2-18. Examples of correct and incorrect focus-control settings.

has a value of 2.25μf, and R3 has a value of 1 megohm, with a resulting
time constant of ¼ second. In turn, the cutoff frequency is defined as
0.6 Hz. However, the input waveform is weakened to 50 per cent amplitude at the cutoff frequency, and the practical low-frequency limit in this
example is approximately 10 Hz.

2.4 BASIC HORIZONTAL-AMPLIFIER CIRCUITRY

Most horizontal amplifiers are designed as paraphase inverters. This
type of amplifier is depicted in Fig. 2-19. Note that a single-ended input
voltage is amplified and split into a double-ended (push-pull) output
voltage. The advantage of double-ended output is that one deflecting plate
in the CRT is driven positive while the other deflecting plate is being driven

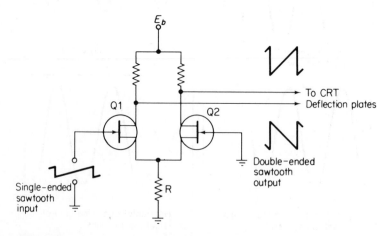

Fig. 2-19. Basic paraphase inverter configuration.

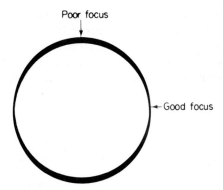

Fig. 2-20. Astigmatism permits partial focus only of a pattern.

negative. In turn, the average potential of both plates is zero at all times. This condition minimizes astigmatism distortion, as exemplified in Fig. 2-20. On the other hand, if a horizontal amplifier provides single-ended output, one of the deflection plates is driven while the other plate is at AC ground potential, as in Fig. 2-16. This condition results in a variation of the average potential of both plates, and causes astigmatism, as shown in Fig. 2-20.

In Fig. 2-19, single-ended input is converted into double-ended output by means of source coupling. That is, when an input signal is applied to the gate of Q1, a corresponding source current flows through R. In turn,

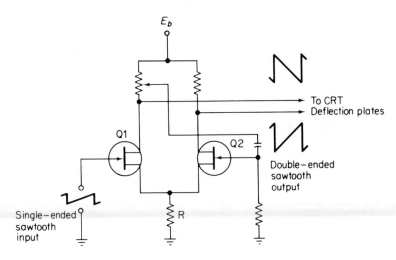

Fig. 2-21. Horizontal-amplifier stage with source coupling and coupling from drain of Q1 to gate of Q2.

the input signal is reproduced across R, and applied to the source of Q2. Accordingly, amplified output signals appear at the drains of Q1 and Q2, and due to the fact that Q2 is source-driven, these output signals appear in push-pull. Since the gain of a paraphase-inverter stage is comparatively small, additional drive is sometimes applied to Q2 by coupling from the drain of Q1, as exemplified in Fig. 2-21. This arrangement permits

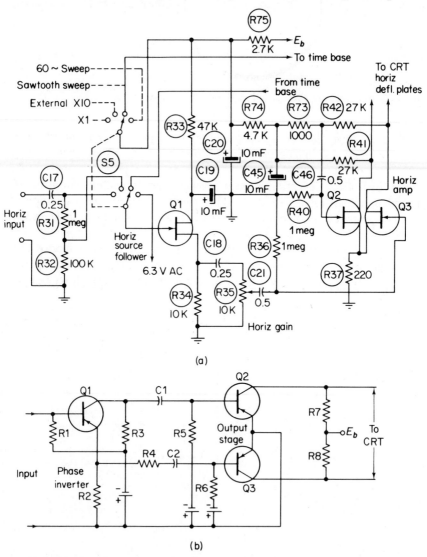

(a)

(b)

Fig. 2-22. Basic horizontal-amplifier circuitry. (a) A complete simple horizontal section. (b) Junction-transistor horizontal-amplifier arrangement.

the value of R to be reduced (decreases the amount of negative feedback), and provides increased amplification by both Q1 and Q2.

Figure 2-22(a) shows a typical horizontal-amplifier section for a simple oscilloscope. It employs a source-follower input stage and a push-pull output stage similar to the configuration shown in Fig. 2-21. A source follower provides a high value of input impedance to the horizontal-amplifier section. This feature is very helpful when one is testing circuits with high internal impedance, as explained in following chapters. Note that the switching arrangement in Fig. 2-22(a) provides for a choice of sawtooth, 60-Hz sine wave, or external drive voltage. Sixty-Hz sine-wave horizontal deflection is used chiefly in visual-alignment procedures, as explained in greater detail subsequently. A voltage-divider circuit comprising R31 and R32 is provided for step attenuation of external drive voltages. Vernier attenuation of sawtooth, 60-Hz, or external drive voltage is provided by R34. Fig. 2-22(b) shows a widely used junction-transistor horizontal-amplifier configuration. It is treated in greater detail in the next chapter.

Note in Fig. 2-22 that the supply voltage to the sawtooth oscillator is switched off when one is operating on 60-Hz sine-wave deflection or external deflection. It is necessary to disable the sawtooth oscillator when it is not in use, to avoid objectionable crosstalk in the horizontal section. The horizontal amplifier exemplified in Fig. 2-22 is AC-coupled throughout. However, other arrangements are DC-coupled. The chief advantage in DC coupling occurs in industrial-electronics servicing when very low frequency waveforms are to be displayed. The high-frequency response of an RC-coupled amplifier is determined chiefly by the values of the drain load resistors (approximately 27 k in Fig. 2-22). In turn, the frequency response extends to about 100 kHz, as seen in Fig. 2-23. Note that the amplifier

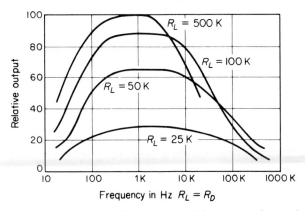

Fig. 2-23. Frequency response of RC-coupled amplifiers for various values of drain load resistance.

gain increases as the drain-load resistance increases, but that the frequency response decreases.

A blanking amplifier is employed to step up the amplitude of the blanking pulse sufficiently to cut off the electron beam in the CRT during retrace. Figure 2-24 exemplifies a blanking-amplifier and free-running time-base configuration. The blanking pulse is taken from the common-source resistor in the sweep-oscillator circuit, and is applied to the gate of Q1. In

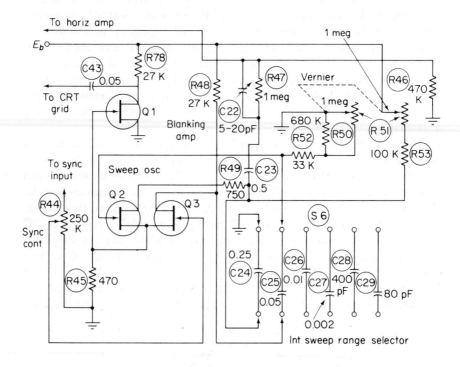

Fig. 2-24. Basic blanking-amplifier and free-running time-base configuration.

turn, the amplitude of the blanking pulse is stepped up by the blanking amplifier and applied to the CRT grid. Note in passing that a trimmer capacitor C22 is provided in the output lead of the sweep oscillator. This is a waveshaping capacitor which is adjusted to provide optimum sweep linearity. Figure 2-25 exemplifies a horizontal control layout for an oscilloscope.

44

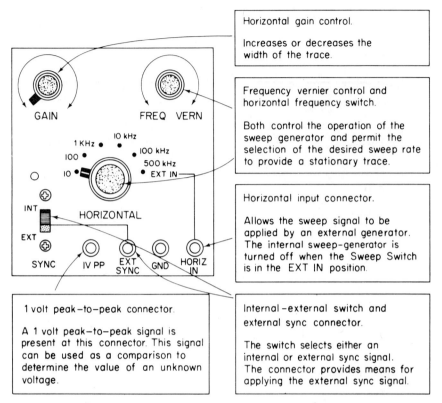

Horizontal gain control.

Increases or decreases the width of the trace.

Frequency vernier control and horizontal frequency switch.

Both control the operation of the sweep generator and permit the selection of the desired sweep rate to provide a stationary trace.

Horizontal input connector.

Allows the sweep signal to be applied by an external generator. The internal sweep-generator is turned off when the Sweep Switch is in the EXT IN position.

1 volt peak-to-peak connector.

A 1 volt peak-to-peak signal is present at this connector. This signal can be used as a comparison to determine the value of an unknown voltage.

Internal-external switch and external sync connector.

The switch selects either an internal or external sync signal. The connector provides means for applying the external sync signal.

GAIN   FREQ VERN

10 kHz
1 KHz • •
100 •   • 100 kHz
             500 kHz
10 • •   • EXT IN

INT

HORIZONTAL

EXT

SYNC   IV PP   EXT   GND   HORIZ
                 SYNC   IN

Fig. 2-25. A horizontal control layout for an oscilloscope. (*Courtesy of* Heath Co.)

2.5 PRINCIPLES OF TRIGGERED TIME BASES

A triggered time base employs a sawtooth oscillator that is biased beyond cutoff, so that the oscillator is not free-running. However, if a trigger pulse is applied to the oscillator, it goes through one cycle of operation and generates a single sawtooth waveform. A triggered time base has two chief advantages in comparison to a free-running time base. First, the period of the sawtooth waveform may be only a small fraction of the period of the waveform that is being displayed. In turn, a small section of a waveform can be picked out and greatly expanded on the CRT screen without a confused pattern such as would result from the use of a free-running time base. Second, a triggered time base can be accurately calibrated for elapsed-time measurements, so that the duration of any part of a displayed waveform can be observed precisely. As a simple illustration of waveform ex-

45

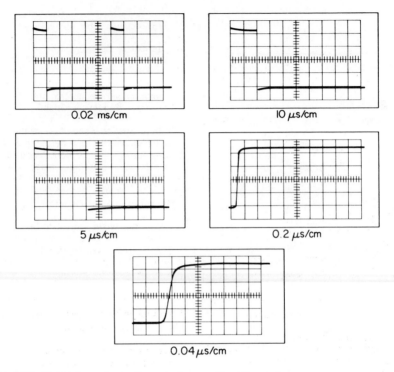

Fig. 2-26. Expansion of a pulse waveform using triggered sweeps.

pansion, Fig. 2-26 shows how the leading edge of a pulse waveform may be expanded progressively by employing sweep rates from 0.2 millisecond per centimeter to 0.04 microsecond per centimeter. This topic will be treated in greater detail subsequently.

A triggered time base is designed around a monostable multivibrator, also termed a one-shot, single-swing, or single-shot multivibrator. Its basic configuration is shown in Fig. 2-27. Since transistor Q1 is biased beyond cutoff, there is no output from the multivibrator in its resting state, and there is no horizontal deflection of the CRT electron beam. On the other hand, when a trigger pulse is applied to the base of Q1, the transistor is momentarily brought out of cutoff and the one-shot multivibrator goes through one complete cycle of operation. Thereby, an output pulse is generated which has a precise duration determined by the time constant of $R_{F1}C_{F1}$. Observe in Fig. 2-27 that Q1 can be triggered only by a negative-going input pulse. This pulse is produced in the trigger section, depicted in Fig. 2-28.

A triggered time base is designed so that a desired portion of a waveform can be selected and displayed on the CRT screen in expanded form.

46

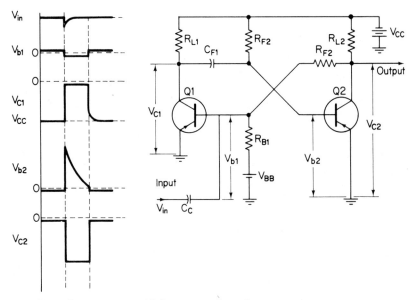

Fig. 2-27. Basic one-shot multivibrator configuration.

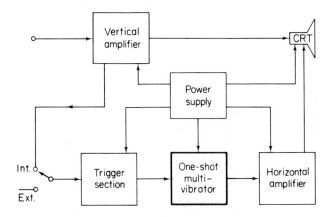

Fig. 2-28. Basic plan of a triggered-sweep oscilloscope.

For example, Fig. 2-29 shows how a waveform may be triggered at a low point or at a high point on its positive slope, or at a low point or at a high point on its negative slope. A trigger pulse is produced by the trigger section only when a sync signal of suitable level is applied. It follows from Fig. 2-29 that the availability of a sync phase inverter in the trigger section will provide a choice of positive-slope or negative-slope triggering. Furthermore, it is seen in Fig. 2-30 that the trigger level can be varied by

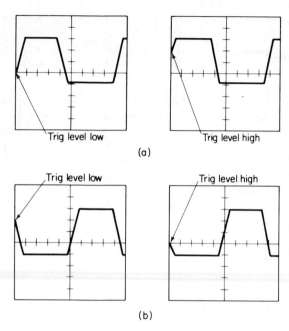

Trig level low

Trig level high

(a)

Trig level low

Trig level high

(b)

Fig. 2-29. Triggering at various points on the leading edge of a waveform. (a) Positive-slope triggering. (b) Negative-slope triggering.

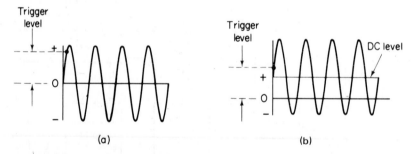

Fig. 2-30. Changing the trigger point on a waveform by addition of a DC component.

adding a suitable amount of DC voltage to the AC sync-signal voltage. In other words, the addition of a DC component serves to raise or lower the sync-signal level with respect to the triggering point.

Figure 2-31 shows a block diagram of a complete triggered-sweep system. Either internal or external sync waveforms may be applied to the sync amplifier. The settings of the sync-amplifier controls determine whether the sweep will be triggered on the positive or negative portion of the sync waveform, and at what point on its leading edge. At this selected point,

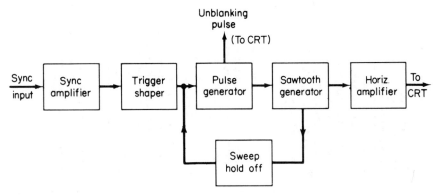

Fig. 2-31. Block diagram of a complete triggered-sweep system.

the trigger shaper produces a sharp trigger pulse. In turn, the pulse generator is triggered and proceeds to apply an unblanking pulse to the CRT and to turn the sawtooth generator on. The unblanking pulse is applied to the CRT grid, and turns the electron beam on for the duration of the sawtooth. This avoids the danger of burning the screen while the CRT is resting.

Note in Fig. 2-31 that the sawtooth generator drives the horizontal amplifier and also energizes the sweep-holdoff circuit. Its function is as follows: When the sawtooth output from the generator reaches a certain level determined by the setting of the sweep-length control, the sawtooth generator turns itself off by means of a turn-off pulse applied by the sweep-holdoff circuit to the pulse generator. In addition, the sweep-holdoff circuit blocks any output from the trigger shaper until after the turn-off pulse arrives. This holdoff action prevents false triggering of the sawtooth generator by trigger pulses other than the one for which the operating controls have been set.

Figure 2-32 exemplifies triggered-sweep operating controls for a typical oscilloscope, with explanations of control functions. Note that in the automatic trigger mode, the oscilloscope operates with a free-running time base, so that a horizontal trace is visible on the CRT screen when no vertical-input signal is present. This free-running automatic sweep has a repetition rate of 100 Hz. However, when a vertical-input signal is present, the automatic triggering rate will speed up, and the sweep rate will equal the repetition rate of the vertical-input signal. Thus, any waveform with a frequency between 100 Hz and 10 MHz (in this example) will be automatically synchronized. The number of cycles displayed on the screen is determined by the settings of the time/cm switch and the horizontal-time variable control.

49

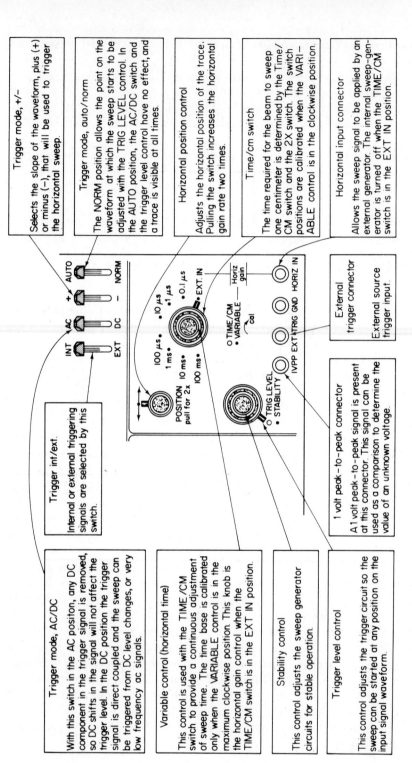

Trigger mode, +/-

Selects the slope of the waveform, plus (+) or minus (−), that will be used to trigger the horizontal sweep.

Trigger mode, auto/norm

The NORM position allows the point on the waveform at which the sweep starts to be adjusted with the TRIG LEVEL control. In the AUTO position, the AC/DC switch and the trigger level control have no effect, and a trace is visible at all times.

Horizontal position control

Adjusts the horizontal position of the trace. Pulling the switch increases the horizontal gain rate two times.

Time/cm switch

The time required for the beam to sweep one centimeter is determined by the Time/CM switch and the 2X switch. The switch positions are calibrated when the VARIABLE control is in the clockwise position.

Horizontal input connector

Allows the sweep signal to be applied by an external generator. The internal sweep-generator is turned off when the TIME/CM switch is in the EXT IN position.

Trigger int/ext.

Internal or external triggering signals are selected by this switch.

External trigger connector

External source trigger input.

1 volt peak-to-peak connector

A 1 volt peak-to-peak signal is present at this connector. This signal can be used as a comparison to determine the value of an unknown voltage.

Trigger mode, AC/DC

With this switch in the AC positon, any DC component in the trigger signal is removed, so DC shifts in the signal will not affect the trigger level. In the DC position the trigger signal is direct coupled and the sweep can be triggered from DC level changes, or very low frequency ac signals.

Variable control (horizontal time)

This control is used with the TIME/CM switch to provide a continuous adjustment of sweep time. The time base is calibrated only when the VARIABLE control is in the maximum clockwise position. This knob is the horizontal gain control when the TIME/CM switch is in the EXT IN position.

Stability control

This control adjusts the sweep generator circuits for stable operation.

Trigger level control

This control adjusts the trigger circuit so the sweep can be started at any position on the input signal waveform.

Fig. 2-32. Typical triggered-sweep operating controls. (*Courtesy of Heath Co.*)

Since a triggered-sweep oscilloscope operates in a similar manner to a free-running oscilloscope when the automatic trigger mode is utilized, it is helpful for the beginner to start his practical experience in this manner. After experience with waveform display is acquired, attention can then be turned to the triggered mode of operation. The operator may be disconcerted by the fact that the screen remains dark unless the triggered-sweep operating controls are suitably adjusted. Therefore, it is often helpful to switch back and forth between the triggered and the automatic modes of operation, to check on the presence of a vertical-input signal. Note that the CRT screen is much more likely to be burned in a triggered-sweep oscilloscope than in a free-running oscilloscope. Accordingly, the operator should advance the intensity control of a triggered-sweep oscilloscope with due caution.

2.6 OPERATION AND TROUBLESHOOTING OF SAWTOOTH OSCILLATORS

In addition to understanding the operation of a sawtooth oscillator, it is also important to become familiar with troubleshooting procedures. Sawtooth oscillators are closely related to square-wave oscillators. Therefore, the student should perform the following square-wave and sawtooth demonstrations and tests, and answer all the questions asked with reference to the various tests and measurements.

Square-Wave Project

SUBJECT

The basic square-wave generator.

OBJECTIVE

To learn how to test a square-wave generator and to determine whether it is operating properly.

MATERIAL REQUIRED

    1. Square-wave generator chassis (Fig. 2-33).

    2. Bench power supply.

    3. Audio oscillator with leads.

    4. Service-type oscilloscope with leads.

    5. VTVM or TVM with leads.

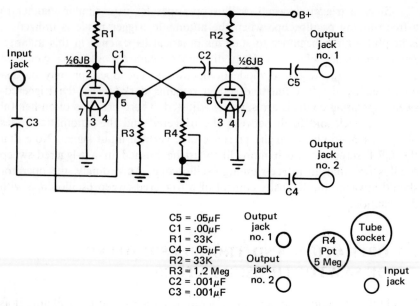

Fig. 2-33. Square-wave generator.

Instructions

Careful attention to this project and good practices in experimental procedures will avoid much future trouble and delay. The following precautions should be observed.

1. Check each connection as you proceed.

2. Always remove the tube when working on the back of an unmounted chassis. Miniature-tube pins bend easily, and the tube may be ruined if not carefully handled.

3. At the completion of this project, make sure that the square-wave generator chassis is left in normal operating condition. Otherwise, you may waste time in the next project before discovering that the square-wave chassis is inoperative or is not operating normally.

Procedure

Testing

1. Assemble the equipment previously noted under Material Required.

2. With the circuits "dead," use the ohmmeter to make the resistance measurements indicated in Table 2-1. Mark down the observed readings in the proper column.

TABLE 2-1 Resistances

OHMMETER CONNECTIONS	REPRESENTATIVE READINGS	OBSERVED READINGS
B plus and plate no. 1 (pin 2)	33,000 ohms	
B plus and plate no. 2 (pin 1)	33,000 ohms	
Grid no. 1 (pin 5) and chassis	1.2 megohms	
Grid no. 2 (pin 6) and chassis	0–5 megohms (depending upon the potentiometer setting)	
Plate no. 1 and grid no. 2 (pins 1 and 6)	Infinity	
Plate no. 2 and grid no. 1 (pins 1 and 5)	Infinity	
INPUT jack and chassis, OUTPUT no. 1 jack and chassis, and OUTPUT no. 2 jack and chassis	Infinity (in each case)	
Cathode (pin 7) and chassis	Zero	
Heaters (pins 3 and 4) and chassis	Infinity	

3. Compare the observed readings with the representative readings listed in the table. If the two sets of readings do not agree reasonably well, try to determine the cause of the discrepancy and correct it. If the trouble cannot be located, consult an instructor.

4. Apply power from the bench power supply to the square-wave chassis, but do not insert the tube into the generator socket at this time.

5. Make voltage measurements at the points indicated in Table 2-2, and insert the observed readings in the proper column.

TABLE 2-2 Voltages

Voltmeter Connections	Representative Readings	Observed Readings
B plus and chassis	180–300 volts, DC	
Plate no. 1 (pin 2) and chassis	Same as B plus	
Plate no. 2 (pin 1) and chassis	Same as B plus	
Grid no. 1 (pin 5) and chassis	Zero	
Grid no. 2 (pin 6) and chassis	Zero	
Cathode (pin 7) and chassis	Zero	
Heaters (pins 3 and 4)	6.3 volts, AC	

6. Compare the observed readings with the representative readings listed in the table. If the two sets of readings do not agree reasonably well, try to determine the cause of the discrepancy and correct it. If the trouble cannot be located, consult an instructor.

Adjustment

1. After correct voltages have been verified in the square-wave generator chassis, insert the 6J6 tube in its socket, and connect the oscilloscope between output no. 1 jack and ground. Set the potentiometer to its mid-position.

2. Adjust the oscilloscope controls to display three cycles of the applied waveform. Draw a sketch of this waveform.

3. Move the vertical-input lead of the oscilloscope to output no. 2 jack (Fig. 2-33). Draw a sketch of the displayed waveform.

4. By comparison with the output signal from an audio oscillator, measure the output frequency of the square-wave generator under the following conditions.

   (a) At the fully clockwise setting of R4 (Fig. 2-33), or as far as the potentiometer can be turned and still maintain oscillation.

   (b) At the fully counterclockwise setting of R4 (Fig. 2-33), or as far as the potentiometer can be turned and still maintain oscillation. Record the frequencies measured under (a) and (b).

5. Measure the peak voltage output of the square-wave generator by means of the oscilloscope, VTVM (or TVM), and the audio oscillator. (See the information sheet on the measurement of non-sinusoidal voltage at the end of this project.) Record the value of the measured peak voltage.

6. Adjust the square-wave generator for production of a symmetrical waveform according to the following procedure.

   (a) Connect the oscilloscope to output no. 2 jack of the generator.

   (b) Adjust the oscilloscope controls to display three cycles of the square waveform.

   (c) Move the pattern up or down on the screen until one of the heavy black horizontal lines of the graticule falls half-way between the top and bottom of the square wave.

   (d) Adjust the potentiometer of the square-wave generator until the width of the positive alternation and the middle square wave equals that of the negative alternation. Adjust the sweep-frequency control of the oscilloscope to retain a three-cycle display of the square waveform on the CRT screen.

   (e) Draw a sketch showing the position of the potentiometer control for production of a symmetrical square waveform.

7. Adjust the square-wave generator for production of asymmetrical square waveforms (rectangular waveforms) according to the following procedure.

   (a) Repeat (b) and (c) of step 6, above.

   (b) Adjust the potentiometer until the positive alternation is as narrow as possible.

   (c) Sketch the position of the potentiometer control for production of this narrow positive pulse. These sketches will be utilized in subsequent experiments.

CONCLUSIONS

1. What effect does varying the square-wave generator potentiometer have upon:

   (a) The frequency of the square wave?
   (b) The amplitude of the square wave?
   (c) The shape of the square wave?

2. Describe any difficulties that you have encountered in this project, and write a report on what measures were taken to correct them. What were the results of your troubleshooting procedures?

2.7 NONSINUSOIDAL WAVEFORM VOLTAGE MEASUREMENTS

This information sheet provides an introduction to the topic of non-sinusoidal waveform voltage measurement. In order to measure the effective (rms) voltage of a sine wave from an audio oscillator, commercial power line, etc., it is necessary only to connect an AC voltmeter across the source and read the effective voltage directly from the meter scale as exemplified in Fig. 2-34(a). If the peak value is desired, instead of the rms value, multiply the meter rms reading by 1.414. To obtain the peak-to-peak amplitude, multiply the rms reading by 2.828 as exemplified in Fig. 2-34(b).

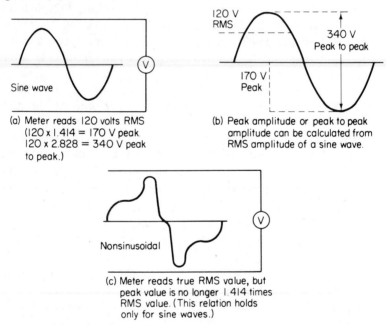

(a) Meter reads 120 volts RMS
(120 x 1.414 = 170 V peak.
120 x 2.828 = 340 V peak
to peak.)

(b) Peak amplitude or peak to peak
amplitude can be calculated from
RMS amplitude of a sine wave.

(c) Meter reads true RMS value, but
peak value is no longer 1.414 times
RMS value. (This relation holds
only for sine waves.)

Fig. 2-34. Curves showing why AC voltmeter cannot be used to measure peak amplitudes of nonsinusoidal waveforms.

On the other hand, voltage measurements of nonsinusoidal waveforms may not be so directly made. As an illustration, most AC meters are calibrated to indicate correct rms values of sine waveforms only. In turn, if the waveform of the voltage under measurement is square, triangular, or some other complex shape, the meter rms reading will be incorrect, and will provide little or no practical information concerning the waveform amplitude, as exemplified in Fig. 2-34(c). Data required for amplitude measurements of complex waveforms can be easily obtained, however, with the aid of an oscilloscope, a voltmeter, and a source of sine-wave voltage. The method followed consists of adjusting the sine-wave amplitude until it produces an oscilloscope pattern of the same height as the distance between the center line and the peak of the pattern produced by the nonsinusoidal waveform. An AC voltmeter is used to measure the rms value of the sine wave. This rms reading is converted to its corresponding peak-to-peak value by multiplying it by 2.828. The resulting peak-to-peak voltage value is the desired answer—it is the amplitude of the nonsinusoidal waveform.

The procedure is as follows:

1. Connect the vertical-input terminals of the oscilloscope across the voltage source to be measured as depicted in Fig. 2-35, and adjust the oscilloscope controls to display one cycle of the waveform, as shown in Fig. 2-36(a).

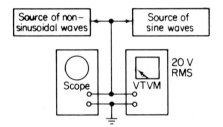

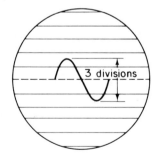

Example: With the sine–wave source adjusted to produce a 3–division deflection, v.t.v.m. reads 20 volts, rms. A–F wave has peak-to-peak amplitude of 20 x 2.828 = 56.56 volts. Peak-to-peak amplitude of nonsinusoidal wave is also 56.56 volts.

Fig. 2-35. Block diagram of equipment used for measuring amplitude of nonsinusoidal waveforms.

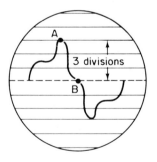

(a) Positive peak is 3 divisions above zero axis

(b) Adjust A–F oscillator output until peaks are 3 divisions apart

Measure rms value of sine wave by placing a v.t.v.m. across a-f oscillator terminals. Multiply observed voltage by 2.828. This gives the voltage represented by a deflection of 3 scope divisions.

Fig. 2-36. Easy method of measuring amplitude of nonsinusoidal waveforms.

2. Adjust the gain controls of the oscilloscope so that the pattern fills most of the screen.

> **Note:** Make no further adjustment of the oscilloscope controls until the measurement is completed; otherwise, the result will be in error.

3. Measure the distance on the screen between the points on the pattern whose *voltage difference* you wish to measure [for example, points *A* and *B* in Fig. 2-36(a)]. This distance can be measured with a ruler, a piece of graph paper, or a graticule over the face of the CRT.

4. Disconnect the voltage source that is being measured, and connect the vertical-input terminals of the oscilloscope to the output terminals of an audio oscillator.

5. Adjust the oscillator frequency control to provide a stationary cycle of sine-wave display on the CRT screen.

6. Adjust the output voltage of the audio oscillator until the peaks are exactly the same distance apart (vertically) as the distance measured in step 3. See Fig. 2-36(b). The peak-to-peak amplitude of the sine wave is now equal to the voltage difference that it is desired to measure.

7. Connect a VTVM or TVM across the audio-oscillator output terminals and read the voltage that is being supplied. Remember that the meter indicates rms voltage values.

8. To convert this rms voltage to its corresponding peak-to-peak value, multiply the meter reading by 2.828.

9. This peak-to-peak amplitude is the voltage existing between the two points (*A* and *B*) on the nonsinusoidal waveform.

Sawtooth Wave Project

SUBJECT

The basic sawtooth waveform generator.

OBJECTIVE

To learn how to test a sawtooth generator and to determine whether it is operating properly.

Material Required

1. Sawtooth generator chassis (Fig. 2-37).

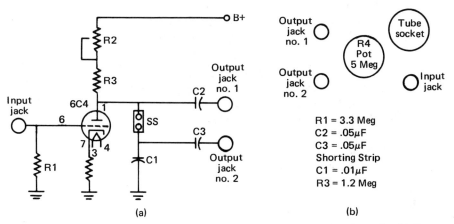

Fig. 2-37. Schematic diagram of sawtooth generator chassis. (a) Schematic. (b) Bottom view.

2. Square-wave generator chassis (Fig. 2-33).
3. Bench power supply.
4. Audio oscillator with leads.
5. Service-type oscilloscope with leads.
6. VTVM or TVM with leads.

Procedure

1. Study the schematic diagram and layout drawing for the sawtooth generator chassis.
2. Do not insert the tube at this time.
3. Connect the sawtooth generator to the power supply, but do not apply power at this time.

Testing

1. Using the ohmmeter, make the resistance measurements indicated in Table 2-3. Mark down your observed readings in the proper column.

TABLE 2-3 Resistances

OHMMETER CONNECTIONS	REPRESENTATIVE READINGS	OBSERVED READINGS
B plus and plate (pins 1 and 5)	1.2 megohms to 6.2 megohms (depending upon potentiometer setting)	
Grid (pin 6) and chassis	3.3. megohms	
Plate (pins 1 and 5) and OUTPUT no. 1 jack	Infinity	
Plate (pins 1 and 5) and OUTPUT no. 2 jack	Infinity	
Cathode (pin 7) and chassis	Zero	
Heaters (pins 3 and 4) and chassis	Infinity	

2. Compare the observed readings with the representative readings listed in the table. If the two sets of readings do not agree reasonably well, try to determine the cause of the discrepancy and correct it. If the trouble cannot be located, consult an instructor.

3. Apply power to the sawtooth generator chassis, and make voltage measurements at the points indicated in Table 2-4. Mark down your observed readings in the proper column.

TABLE 2-4 Voltages

VOLTMETER CONNECTIONS	REPRESENTATIVE READINGS	OBSERVED READINGS
B plus and chassis	200–300 volts, DC	
Plate (pins 1 and 5) and chassis	Same as B plus	
Grid (pin 6) and chassis	Zero	
Cathode (pin 7) and chassis	Zero	
Heaters (pins 3 and 4)	6.3 volts, AC	

4. Compare the observed readings with the representative readings listed in the table. If the two sets of readings do not agree reasonably well, try to determine the cause of the discrepancy and correct it. If the trouble cannot be located, consult an instructor.

Adjustment

1. Insert the 6C4 tube in its socket of the sawtooth generator. Set up the square-wave generator, the sawtooth generator, and the oscilloscope as shown in the block diagram of Fig. 2-38.

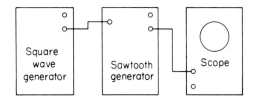

Fig. 2-38. Block diagram of equipment setup for adjustment of sawtooth generator.

2. Set the potentiometer of the square-wave generator to obtain wide negative pulses.
3. Connect the oscilloscope to output no. 2 jack on the square-wave generator chassis, and check to make certain that the required waveform is being generated.
4. Return the oscilloscope input lead to the output of the sawtooth generator, and adjust the oscilloscope to display three cycles of the output waveform on the CRT screen. Either output no. 1 or no. 2 jack may be used.
5. Measure the peak-to-peak voltage of the sawtooth generator output waveform, employing the procedure that you learned previously. Record this measured value.

Study of Waveforms Available

1. With the square-wave-generator potentiometer set to produce a wide negative pulse, and with the sawtooth-generator potentiometer set at its mid-position, carefully observe the waveform displayed on the CRT screen. Draw to scale two cycles of this waveform, in the space provided at *B* in Table 2-5.

TABLE 2-5 Waveforms Available from Sawtooth Generator

A
Wide Negative Input Pulses, Narrow
Positive Spikes, Potentiometer
Counterclockwise

D
Symmetrical Square-Wave Input,
Potentiometer Counterclockwise

B
Wide Negative Input Pulses, Narrow
Positive Spikes, Potentiometer
at Mid-scale

E
Symmetrical Square-Wave Input,
Potentiometer at Mid-scale

C
Wide Negative Input Pulses, Narrow
Positive Spikes, Potentiometer
Clockwise

F
Symmetrical Square-Wave Input,
Potentiometer Clockwise

2. Turn the sawtooth-generator potentiometer as far clockwise as possible without stopping oscillation, and trace the resulting waveform at A in Table 2-5.

3. Turn the potentiometer as far clockwise as possible without stopping oscillation, and trace the resulting waveform at C in Table 2-5.

4. Now set the square-wave-generator potentiometer for approximately symmetrical square-wave output. Make a fine adjustment of the potentiometer until the output wave is as nearly symmetrical as possible.

5. Reconnect the oscilloscope input lead to one of the output jacks on the sawtooth generator, and observe the waveform that is displayed.

6. At D, E, and F in Table 2-5, draw to scale two cycles of the waveform obtained with the sawtooth-generator potentiometer in its counterclockwise, mid-point, and clockwise positions, respectively.

CONCLUSIONS

1. What effect does varying the sawtooth-generator potentiometer have upon:

   (a) The frequency of the sawtooth waveform?
   (b) The amplitude of the sawtooth waveform?
   (c) The shape of the sawtooth waveform?

2. What effect does changing the width of the input square wave have upon the shape of the sawtooth? Does a narrow spike or a broad, symmetrical square-wave input produce greater linearity in the rising portion of the sawtooth?

3. Describe any difficulties encountered in making the sawtooth generator operate properly. What measures were taken to overcome these difficulties, and what were the results?

3

VERTICAL AMPLIFIERS

3.1 GENERAL CONSIDERATIONS

Most oscilloscope applications require vertical amplifiers, because the available test signal voltage is generally insufficient to produce adequate vertical deflection of the CRT beam. Vertical amplifiers are classified into narrow-band and wide-band types, with AC response only, or with both AC and DC response. Narrow-band and wide-band amplifiers are not defined with respect to exact frequency response. However, an amplifier that covers the audio-frequency range from 20 Hz to 20 kHz is in the narrow-band category. Amplifiers for certain industrial-type oscilloscopes may cover less than the audio-frequency range. An amplifier with frequency response to 2 MHz may be classified as a medium-band type. Again, an amplifier with response to 4.5 MHz is usually termed as wide-band type.

However, these terms are relative, and as an illustration, a sophisticated oscilloscope with vertical-amplifier response to 30 MHz is a wide-band instrument in comparison to a color-TV oscilloscope with response to 4.5 MHz. We shall find that narrow-band vertical amplifiers generally have high gain, whereas wide-band amplifiers are limited in respect to maximum available gain. The chief limitation in design of high-gain wide-band amplifiers is the random noise level that is encountered. That is, the random-noise output of an amplifier increases as its bandwidth is increased. Although various design measures can be employed to minimize the noise voltages in a vertical amplifier, there are limits imposed by the present state of the art with respect to the gain-bandwidth figure of a vertical amplifier.

3.2 BASIC VERTICAL-AMPLIFIER CIRCUITRY

High input impedance is a basic vertical-amplifier requirement, in order to minimize loading of the circuit under test. Various input-stage circuits may be employed to obtain a fairly high input impedance. For example, three junction-transistor configurations are depicted in Fig. 3-1, which provide higher input impedance than if the input signal were applied directly to the base and emitter. A common-collector (emitter-follower) circuit is shown in (a). It provides a high input impedance because a large amount of negative feedback occurs in the emitter branch. If R_L is

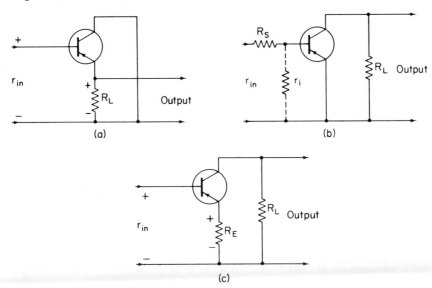

Fig. 3-1. Three junction-transistor amplifier configurations for obtaining a comparatively high input impedance. (a) Common-collector (emitter-follower) circuit. (b) Common-emitter circuit with series base resistor. (c) Common-emitter circuit with emitter degeneration.

65

500 ohms, the input resistance will be greater than 20,000 ohms, as seen in Fig. 3-2. The chief disadvantage of the CC configuration is the variation of input impedance which results with changes in collector current.

Next, Fig. 3-1(b) shows a common-emitter circuit with a series base resistor R_s. As seen in Fig. 3-2, the stage would have an input impedance of about 2000 ohms for a load resistance of 500 ohms, if a series resistor were not inserted. If R_s is 18,000 ohms, the total input resistance becomes 20,000 ohms. For practical purposes, it can be said that the value of input resistance is equal to the value of the series resistor. This circuit

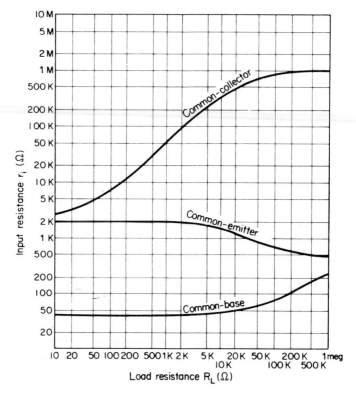

Fig. 3-2. Input resistances for transistors in the CB, CE, and CC configurations.

provides a reasonably constant value of input resistance with changes in collector current. Its chief disadvantage is bias instability with temperature changes, if the bias voltage is fed to the base through R_s. A minor disadvantage is some loss in stage gain due to signal-voltage drop across R_s.

Figure 3-1(c) depicts a common-emitter circuit with emitter degeneration. With an unbypassed emitter resistor of 500 ohms and a load resistor of 500 ohms, the input resistance of the transistor is approximately 20,000 ohms. Although the input resistance is the same as in the foregoing arrangements, emitter degeneration has the advantages of bias stabilization and of greater immunity of input-resistance variation with changes in collector current. Figure 3-3 lists the principal characteristics for junction transistors in the three basic configurations. When emitter degeneration is utilized in the CE configuration, the gain figures are decreased, the input resistance is increased and the output resistance is increased also.

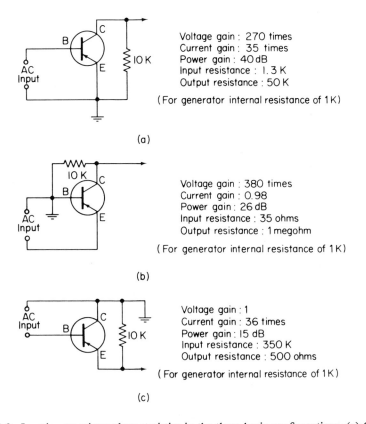

Voltage gain : 270 times
Current gain : 35 times
Power gain : 40 dB
Input resistance : 1.3 K
Output resistance : 50 K

(For generator internal resistance of 1 K)

(a)

Voltage gain : 380 times
Current gain : 0.98
Power gain : 26 dB
Input resistance : 35 ohms
Output resistance : 1 megohm

(For generator internal resistance of 1 K)

(b)

Voltage gain : 1
Current gain : 36 times
Power gain : 15 dB
Input resistance : 350 K
Output resistance : 500 ohms

(For generator internal resistance of 1 K)

(c)

Fig. 3-3. Junction-transistor characteristics in the three basic configurations. (a) Common emitter. (b) Common base. (c) Common collector (emitter follower).

Extremely high input impedance is obtained by means of field-effect transistors for vertical-amplifier input stages. Thus, the arrangement depicted in Fig. 3-4 provides practically infinite impedance over the audio-frequency and video-frequency ranges. A field-effect transistor can be compared with an electron tube in this respect. An FET input stage may drive an FET second stage, or a junction-transistor second stage. In any

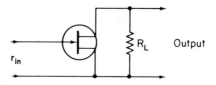

Fig. 3-4. Field-effect transistor amplifier configuration.

event, a third stage is customarily a junction transistor. Transistors may be direct-coupled, as shown in the basic configuration of Fig. 3-5. Note that the base-emitter bias on Q1 is low, compared with the base-emitter bias on Q2. The input stage is operated at low current and low collector voltage to minimize random noise. Since the signal-to-noise ratio is practically determined by the first stage, the second stage is operated at comparatively high current and high collector voltage.

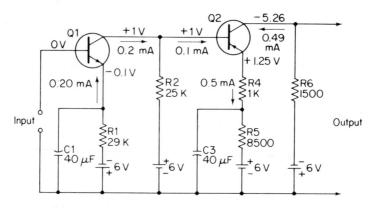

Fig. 3-5. Two-stage direct-coupled preamplifier arrangement.

Figure 3-6 exemplifies a three-stage RC-coupled preamplifier arrangement. An input impedance of approximately 55,000 ohms is obtained by means of R1, the 1500-ohm unbypassed emittter resistor for Q1. Large coupling capacitors are required for good low-frequency response, owing to the comparatively low input impedances of Q2 and Q3. Similarly, large emitter bypass capacitors are required. The chief advan-

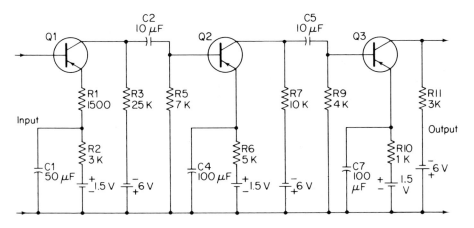

Fig. 3-6. Three-stage RC-coupled preamplifier arrangement.

tage of an AC-coupled configuration is its comparative stability under temperature changes, without elaborated biasing networks. Note that although separate battery voltage sources are shown for the emitter and collector circuits of individual transistors in Figs. 3-5 and 3-6, commercial circuitry employs a common voltage source with various RC decoupling circuits to avoid undesirable feedback.

Next, Fig. 3-7 depicts a one-stage phase-inverter configuration that converts a single-ended input signal into a push-pull output signal. Since the collector signal is 180 deg out of phase with the emitter signal, Q1 drives Q2 and Q3 in push-pull. Equal drive voltages are applied to

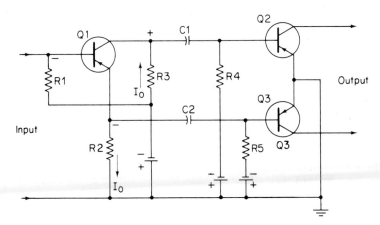

Fig. 3-7. A one-stage phase-inverter configuration.

the bases of Q2 and Q3, because R2 and R3 have the same value. However, the amplitude of drive that can be employed without incurring distortion is somewhat limited, since the emitter impedance of Q1 is considerably lower than its collector impedance. These unequal internal
impedances produce circuit unbalance at high drive levels.

To obtain balanced operation at high drive levels, a resistor may
be inserted in the emitter side of the drive circuit, as seen in Fig. 3-8.
Thus, R4 is chosen with a value that makes the emitter internal impedance of Q1 equal to its collector impedance. In turn, R2 has a somewhat higher value than R3, to compensate for the voltage drop across
R4. Because of the large negative-feedback voltage drop across R2 in
Figs. 3-7 and 3-8, a fairly high amplitude input signal voltage must be
applied to Q1. If more gain is needed, two-stage phase inverters are
utilized. A two-stage phase inverter also provides more power output, and
this feature is sometimes needed in wide-band amplifiers.

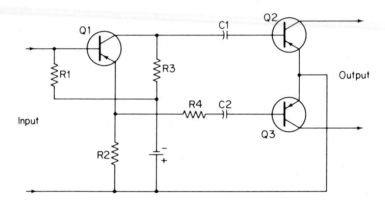

Fig. 3-8. A one-stage phase-inverter configuration with equalized internal impedances.

Figure 3-9 shows a two-stage phase inverter using one CE configuration (Q1) and one CB configuration (Q2), driving a push-pull output stage (Q3 and Q4). This circuit is basically similar to that of Fig. 3-8,
since a CB configuration does not reverse the signal phase from emitter
to collector. Higher gain is provided, due to the contribution of Q2
in Fig. 3-9, and also because the emitter input resistance of a CB stage
is very low. In turn, Q1 effectively has less degeneration in Fig. 3-9 than
in Fig. 3-8. Less degeneration corresponds to higher stage gain. Note that
the phase-inverter circuits that have been shown are AC-coupled; DC-
coupled versions may be employed, if desired.

Next, Fig. 3-10 shows a two-stage phase inverter using two CE

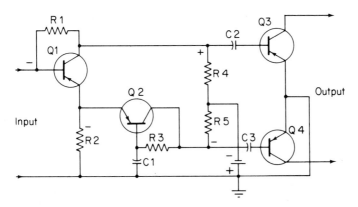

Fig. 3-9. A two-stage phase inverter using CE and CB configurations.

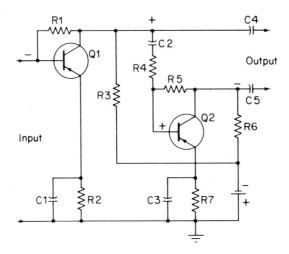

Fig. 3-10. A two-stage phase inverter with two CE configurations.

configurations. An output is taken from the collector of Q1, which also drives the base of Q2. Resistor R4 equalizes the base-drive voltages for Q2 and Q1. R1 and R5 are base biasing resistors. Emitter resistors R2 and R7 are bypassed. However, the emitter resistors provide DC degeneration, thereby contributing to bias stability. This is an RC-coupled inverter configuration, although a DC-coupled counterpart may be utilized in a DC oscilloscope. Since identical CE configurations are used in Fig. 3-10, the output impedances of both sections are equal. Considerably greater output power can be supplied than by a one-stage phase inverter such as shown in Fig. 3-8.

3.3 COMPLETE VERTICAL-AMPLIFIER CONFIGURATION

Figure 3-11 shows a complete vertical-amplifier circuit. The vertical-input signal is coupled through R1 and C1 to the gate of Q1. R1 serves as a protective resistor to prevent damage to Q1 in case of overload. Note that D1 and D2 are transistors connected to provide zener action. They limit the gate voltage on Q1 to ±9 volts. C1 provides improved high-frequency response around R1. Q1 operates as a source follower and provides very high input impedance. Q2 operates as a constant-current source for Q1, thereby stabilizing its operation. D4 and D5 each provide a 0.6 volt drop (total 1.2 volts) and hold the base of Q2 at a constant voltage. Since the circuit for Q2 is basically an emitter-follower configuration, and the emitter voltage is dependent upon the base voltage, the emitter voltage also remains constant. It turn, the current through R2 is constant. R2 is adjusted so that the source voltage of Q1 is zero when no vertical-input signal is applied.

A signal applied to the gate of Q1 will cause only voltage changes at the source because the current through Q1 is constant. These voltage variations are applied across the gain control R404, and a portion of this signal is applied to the gate for the source follower Q3. Q4 provides a constant-current source for Q5 and Q6. Since the emitter of each transistor is connected to the constant-current source, the current source serves as a common-emitter resistance and fixes the operating point for the following stages. The output from source-follower Q3 is amplified by Q5. A portion of the signal applied to the base of Q5 appears at its emitter. Because transistors Q5 and Q6 have a common-emitter resistance, the signal present at the emitter of Q5 is effectively coupled to the emitter of Q6.

Transistor Q6 functions as a common-base amplifier, with its base potential determined by the setting of the vertical-position control R406. This control positions the trace by applying a DC voltage to the base of Q6, thereby establishing a certain DC unbalance in the vertical-amplifier system. When the collector output voltage of Q5 decreases, its emitter voltage increases. In turn, the forward bias on Q6 decreases and its collector voltage increases. The signal at the collector of Q6 is 180 deg out of phase with the signal at the collector of Q5, thereby providing push-pull amplification. C3 is an emitter partial-bypass capacitor that acts to increase emitter degeneration at low frequencies, thereby providing improved high-frequency amplification. R8 and R9 establish the DC gain of the amplifier. Driver transistors Q7 and Q8 operate as common-emitter amplifiers. They provide gain and also isolate Q5 and Q6 from the changing current demand of the output transistors. C4 is an emitter partial-

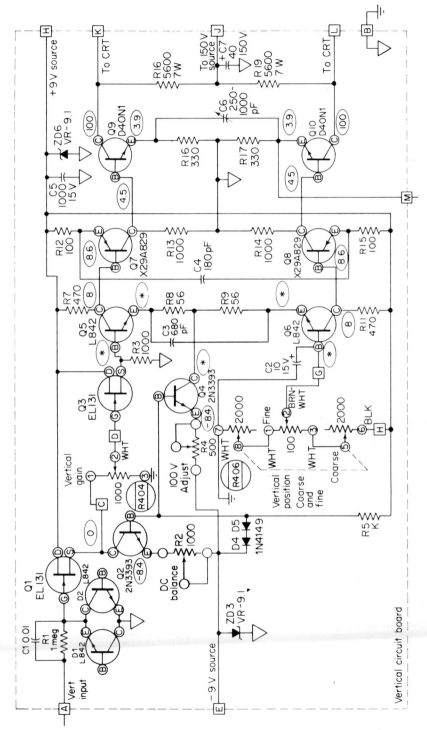

Fig. 3-11. A complete vertical-amplifier circuit. (*Courtesy of* Heath Co.)

bypass capacitor that improves high-frequency response. The output transistors Q9 and Q10 provide final amplification and drive the vertical-
deflecting plates of the CRT.

3.4 VERTICAL ATTENUATOR

Since a very wide range of input signal amplitudes must be accommodated by a vertical amplifier, a vertical-input attenuator is required. Figure
3-12 shows two basic attenuator circuits that operate on the principle
of potentiometer gain control. Continuous attenuation operates satisfactorily, provided that its resistance is low. However, an oscilloscope usually has a vertical-input resistance of at least one megohm. In turn, a
simple attenuator imposes serious distortion on complex waveforms.

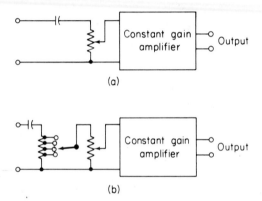

Fig. 3-12. Simple vertical-attenuator arrangements. (a) Continuous control. (b) Step
control.

Figure 3-13 exemplifies square-wave distortions occurring at various settings
of a high-resistance potentiometer. These distortions are caused by the
distributed and stray capacitances in the potentiometer construction.

To obtain signal attenuation without distortion, a compensated step
attenuator is utilized, as shown in Fig. 3-14. Two attenuation positions
are provided. In position no. 1, the vertical-input signal is fed directly
through the switch to the following vertical amplifier. No attenuation of
the input signal occurs. On the other hand, in position no. 2, the vertical-
input signal flows through a voltage divider comprising R1C1 and R2C2.
A signal attenuation of 10 to 1 is generally provided in position no. 2.

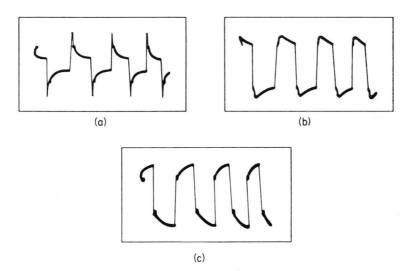

Fig. 3-13. Square-wave distortion imposed by a high-resistance potentiometer. (a) At a low setting. (b) At a mid-range setting. (c) At a high setting. (*Courtesy of* U.S. Armed Forces)

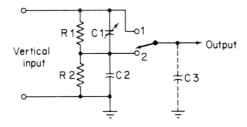

Fig. 3-14. A simple compensated vertical step attenuator.

Note that the sum of the resistances of R1 and R2 is typically 1 megohm. If the resistance of R1 is 0.9 megohm and the resistance of R2 is 0.1 megohm, a vertical-input signal will be attenuated to 0.1 of its original value in position no. 2.

Note in Fig. 3-14 that R1 and R2 provide 10-to-1 attenuation of low signal frequencies or DC, whereas C1 and C2 provide 10-to-1 attenuation of high signal frequencies. The effect of the stray capacitance C3 on the switching circuit is compensated by adjustment of trimmer capacitor C1. Capacitor C2 is included so that adjustment of C1 will bring the time constant R1C1 equal to the time constant R2C2. When C1 is adjusted correctly, the attenuator is said to be compensated, and a complex waveform such as a square wave will be passed without distortion. Misadjustment of C1 becomes evident as the square-wave distortions shown in Fig. 3-15. A square-wave repetition rate in the range from 15 kHz to 25 kHz is generally used to check attenuator compensation.

C1 too large   C1 too small   C1 correct

Fig. 3-15. Square-wave distortions produced by misadjusted compensating capacitor.

Figure 3-16 depicts vertical step-attenuator circuitry for a service-type oscilloscope. Three steps of attenuation are provided, designated as X1, X10, and X100. Two trimmer capacitors, C402 and C405, are utilized for compensation adjustments. The effect on display of a sine wave by vertical step attenuation is exemplified in Fig. 3-17. Fine (vernier) at-

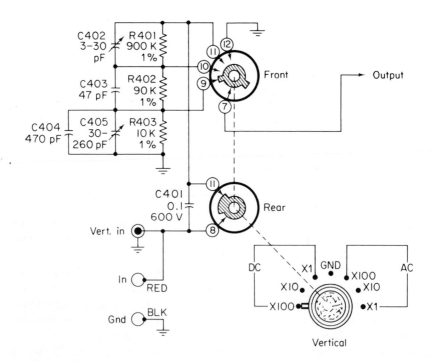

Fig. 3-16. Vertical step-attenuator circuitry for a service-type oscilloscope. (*Courtesy of* Heath Co.)

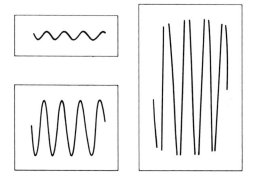

Fig. 3-17. Vertical step attenuation of a sine-wave signal. (*Courtesy of* U.S. Armed Forces)

tenuation of vertical gain is provided by R404 in the example of Fig. 3-11. Note that the potentiometer is energized by a source follower, and has a low value of resistance (1000 ohms). In turn, the effects of stray capacitances are negligible, and the fine attenuator does not require compensation.

Note the coupling capacitor C401 in Fig. 3-16. It provides capacitive coupling to the vertical amplifier on the AC range of the attenuator, and provides direct coupling on the DC range of the attenuator. When DC operation is employed, the coupling capacitor is short-circuited by switching action. The difference between AC and DC operation is exemplified in Fig. 3-18. When an oscilloscope is operated on its AC function, an AC waveform is always displayed with respect to the zero-volt level (resting position of the electron beam on the CRT screen). On the other hand,

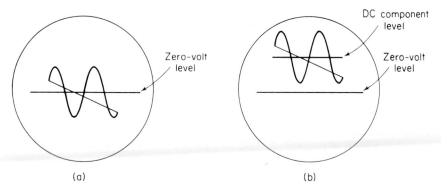

Fig. 3-18. AC and DC displays of a sine waveform. (a) Display on AC function of scope. (b) Display on DC function of scope.

when an oscilloscope is operated on its DC function, an AC waveform will be displayed with respect to the DC component level (providing, of course, that the AC waveform has a DC component).

Figure 3-19 illustrates the distinctions between AC waveforms, pulsating-DC waveforms, and AC with DC component waveforms. Note that an AC waveform has no DC component, and its average value is zero. On the other hand, a pulsating-DC waveform has a DC component, and the AC component does not cross the zero-volt level. Again, an AC

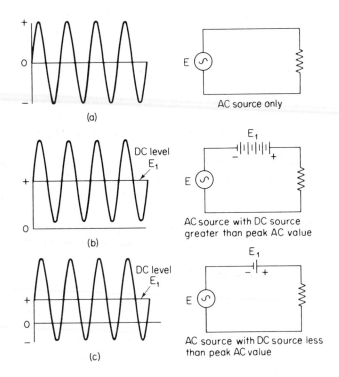

Fig. 3-19. (a) AC waveforms. (b) Pulsating-DC waveforms. (c) AC with DC component waveforms.

waveform with a DC component has an AC component that crosses the zero-volt level. A common example of a pulsating DC waveform is found at the collector of a transistor operating in class *A*. Again, a common example of an AC with DC component waveform is found at the collector of a transistor operating in class *AB*.

3.5 OSCILLOSCOPE PROBES

A probe is basically an input coupling device for application of a signal voltage from a circuit under test to the vertical-input terminals of an oscilloscope. Elaborated probes have signal-processing functions, in addition to their coupling function. Note that an oscilloscope may be used with a pair of vertical-input test leads. Open leads are satisfactory for tests in low-impedance circuits, such as power supplies, but they are unsatisfactory for tests in high-impedance circuits or comparatively high-frequency circuits. An oscilloscope has a typical input resistance and capacitance of 1 megohm and 20 pf at its vertical-input terminal. This high value of input impedance can lead to practical problems when transferred to a circuit under test with a pair of ordinary test leads.

We shall find that a pair of open test leads will pick up any stray fields that happen to be present in their vicinity. Thus, 60-Hz hum fields are invariably present, owing to the building wiring system. Again, if a TV receiver is operating in the same room, 15,750-Hz stray fields will be present. If a high-impedance circuit is being tested with open leads, stray field pickup will combine with the circuit waveform under test, and distort the pattern displayed on the CRT screen. Therefore, it is common practice to use a shielded vertical-input cable with an oscilloscope. A *direct probe* consists simply of a coaxial cable approximately 3.5 feet in length. Although a direct-probe arrangement shields against stray fields, it will load high-impedance circuits objectionably, and produce waveform distortion.

Circuit loading by a direct probe occurs because of its comparatively high input capacitance. As an illustration, the coaxial cable might have an input capacitance of 70 pf. With an oscilloscope input capacitance of 20 pf, a total capacitance of 90 pf will be shunted across the circuit under test. A capacitance of 90 pf will upset the action of high-impedance circuits, some medium-impedance circuits, and various tuned circuits. For this reason, most tests are made with a low-capacitance probe, such as that depicted in Fig. 3-20. Note that the probe circuit bears a marked resemblance to an RC section of a vertical step attenuator. A low-

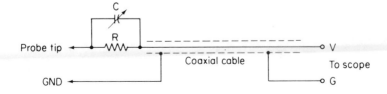

Fig. 3-20. A low-capacitance probe configuration.

capacitance probe is designed to match the oscilloscope with which it is to be used. It is also designed to provide a signal attenuation of 10 to 1.

For example, the input resistance and capacitance to the coaxial cable when connected to the vertical-input terminal of an oscilloscope might be 1 megohm shunted by 80 pf. In such a case, the low-capacitance probe would consist of a 9-pf capacitor C, approximately, and a 9-megohm resistor R. To obtain exact compensation, C is made adjustable. When the time constant of the low-C probe is exactly equal to the time constant of the vertical-input circuit for the oscilloscope, a square wave or other complex waveform will be passed without distortion. There is a "trade-off" involved in low-C probe operation, in that the input capacitance is reduced to $\frac{1}{10}$ of its original value, but reduces the signal strength to $\frac{1}{10}$ of its source value.

Service-type oscilloscopes are generally provided also with a *demodulator probe,* also called a detector probe. This is an example of a signal-processing probe which demodulates an AM signal before it is applied to the oscilloscope. Figure 3-21 shows a typical circuit arrangement and characteristics of a demodulator probe. The usefulness of a demodulator probe is its extension of the effective high-frequency response of an oscilloscope into the very-high frequency range. That is, the demodulator probe develops the modulation envelope of an AM signal. Since the

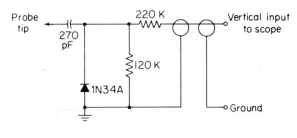

Frequency response characteristics:
```
  RF carrier range . . . . . . . . . . . 500 kHz to 250 MHz
  Modulated-signal range . . . . . . .    30 to 5000 hertz
Input capacitance (approx.). . . . . . .            2.25 pF
Equivalent input resistance (approx.):
     At 500 kHz . . . . . . . . . . . . . . . . 25,000 ohms
          1 MHz . . . . . . . . . . . . . . . . 23,000 ohms
          5 MHz . . . . . . . . . . . . . . . . 21,000 ohms
         10 MHz . . . . . . . . . . . . . . . . 18,000 ohms
         50 MHz . . . . . . . . . . . . . . . . 10,000 ohms
        100 MHz . . . . . . . . . . . . . . . .  5000 ohms
        150 MHz . . . . . . . . . . . . . . . .  4500 ohms
        200 MHz . . . . . . . . . . . . . . . .  2500 ohms
Maximum input:

AC voltage       . . . . . . . . . . . . . . . .  28 peak volts
```

Fig. 3-21. Circuit and characteristics for a demodulator probe.

modulation envelope has a much lower frequency than the carrier component of the waveform, practical tests can be made in VHF circuits.

As an example of demodulator-probe application, it permits signal-tracing procedures in the IF section of a television receiver. The carrier frequency in this section is approximately 43 MHz. When an AM IF waveform is demodulated by a demodulator probe as depicted in Fig. 3-22, the envelope has frequencies extending only to 4 MHz. In turn, this frequency range can be accommodated by the vertical amplifier of a service-type oscilloscope. Note that the input impedance of a simple demodulator probe is comparatively low, and circuit loading is generally encountered. Although waveform distortion occurs, signal tracing procedures are of practical utility to determine whether an IF signal is present or absent at the test point.

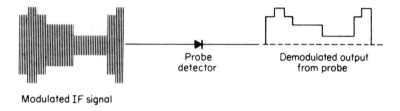

Modulated IF signal

Fig. 3-22. Principle of demodulator-probe action.

4

BASIC
OSCILLOSCOPE
APPLICATIONS

4.1 GENERAL CONSIDERATIONS

Basic oscilloscope applications can be classified into frequency measurements, time measurements, voltage measurements, phase measurements, distortion analysis, transient analysis, and waveform analysis. As an illustration of frequency measurement, Fig. 4-1 depicts a sine-wave display that crosses the horizontal axis on the screen graticule at 2-centimeter intervals. The time base is set for a sweep speed of 100 microseconds per centimeter. In turn, the period of the displayed sine wave is 400 microseconds, which corresponds to a frequency of 2.5 kHz.

Figure 4-2 shows an example of time measurement. The rise time of the leading edge of a waveform is defined as the elapsed time from 10 to 90 per cent of its total amplitude. This definition eliminates unrelated cornering effects from the rise-time value. In the example of Fig. 4-2, the leading edge of a square wave has been expanded horizontally on the screen, so that the rise-time interval is clearly apparent. This is the interval T. If the time base is set for a sweep speed of 0.05 microsecond per centimeter,

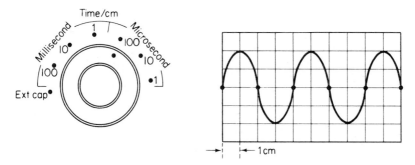

Fig. 4-1. Example of frequency measurement.

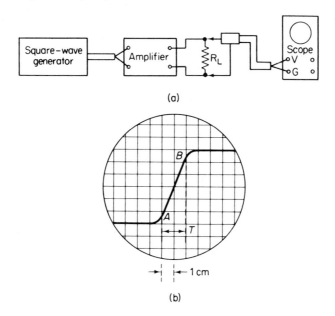

Fig. 4-2. Rise-time measurement of leading edge of square wave.

the rise time is equal to 0.1 microsecond. Other types of time measurements are explained subsequently.

An example of peak-to-peak voltage measurement is depicted in Fig. 4-3. Note that the vertical step attenuator is set to its 20 volts p-p/cm position. In turn, the displayed sine wave has an amplitude of 4 cm, or a voltage of 80 volts p-p. Voltage measurements are also related to decibel measurements, as seen in Fig. 4-4. In this example, a frequency response curve is displayed on the CRT screen. Note that dB values are related to the amplitude of the waveform. That is, the maximum amplitude corre-

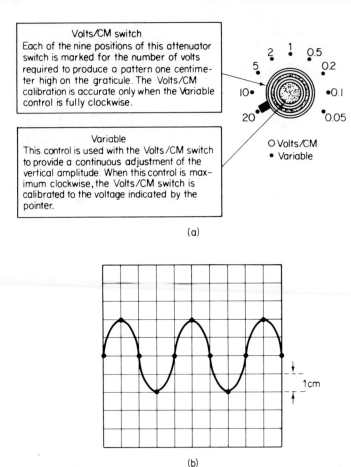

(a)

(b)

Fig. 4-3. Peak-to-peak voltage measurement of a sine waveform. (a) Calibrated vertical step attenuator. (b) Display of 80 volts peak-to-peak. (*Courtesy of* Heath Co.)

sponds to 0 dB, half amplitude corresponds to −6 dB, and zero amplitude corresponds to minus infinity dB. Test setups for displaying frequency response curves are explained in a following chapter.

Previous note has been made of phase measurements. The basic method of measuring the phase difference between two sine-wave voltages is calculated from the Lissajous figure that they display on the CRT screen, as shown in Fig. 4-5. Note that the pattern must be exactly centered on the vertical and horizontal axes, in order to measure the intervals *A* and *B* accurately. This method is useful only for sine waves, and cannot be used with complex waves. If the sine waves have harmonics,

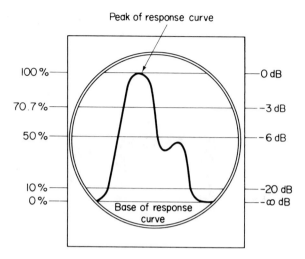

Fig. 4-4. Decibel measurements in a frequency response waveform.

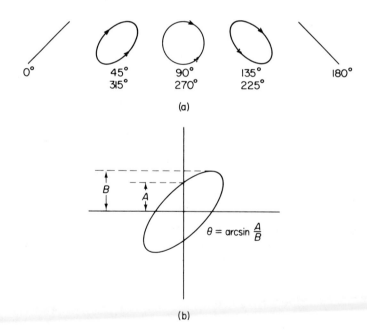

Fig. 4-5. Measurement of phase angle between two sine-wave voltages. (a) Lissajous figures for phase angles at 45-deg intervals. (b) General phase angle measurement.

the elliptical pattern will be distorted and the phase angle cannot be measured accurately. However, a dual-trace oscilloscope such as that illustrated in Fig. 4-6 permits measurement of phase relations between complex waveforms.

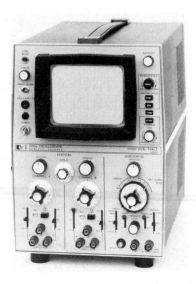

Fig. 4-6. A typical dual-trace oscilloscope. (*Courtesy of* Sencore)

A dual-trace oscilloscope provides two vertical-input channels, so that a pair of waveforms can be applied to the vertical section and displayed one above the other, as exemplified in Fig. 4-7(a). Each vertical channel has its individual step and vernier attenuators and vertical-position control. In the waveform display of Fig. 4-7(b), the vertical-input signals are from different branches in a multivibrator circuit. The negative peak of the upper waveform occurs at the same instant as the positive peak of the lower waveform. There is approximately a 135 deg phase difference between the two waveforms. That is, there are 360 deg from one positive peak to the next positive peak in the upper waveform (and lower waveform). In turn, there are about 135 deg between the positive peak of the upper waveform and the positive peak of the lower waveform. Dual-trace oscilloscopes are considered in greater detail subsequently.

A basic method of distortion analysis is depicted in Fig. 4-8. This method is often used in checking audio amplifiers. Note that the oscilloscope must have highly linear vertical and horizontal amplifiers. Otherwise, distortion in the oscilloscope amplifiers would be falsely charged

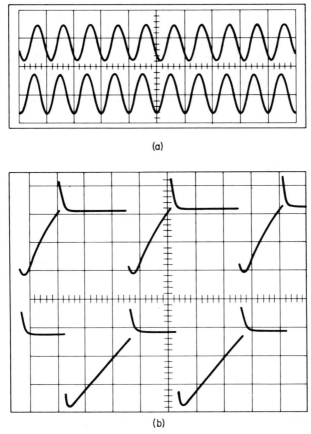

(a)

(b)

Fig. 4-7. Dual-trace displays of typical waveforms. (a) Phase comparison of two sine waves. (b) Phase comparison of two complex waves. (*Courtesy of* Sencore)

to the amplifier under test. The amplifier must be connected to a suitable load resistor, and the output from the audio oscillator is advanced to provide maximum rated power output from the amplifier. It is customary to test an audio amplifier at 1 kHz, although additional tests at lower and higher frequencies can provide supplementary data. If the amplifier is precisely linear, a straight diagonal line is displayed on the CRT screen. Overloading is indicated by curving at the end(s) of the line. Phase shift between input and output of the amplifier is indicated by an elliptical pattern. Amplitude nonlinearity is displayed as a curved line, instead of a straight line. Crossover distortion is indicated by a "jog" at the center of the diagonal line.

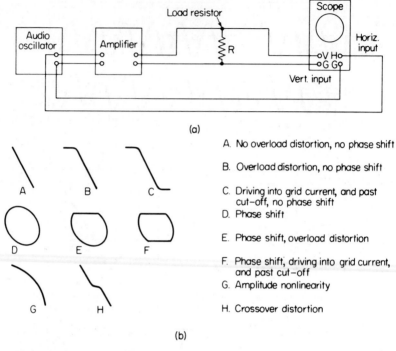

(a)

A. No overload distortion, no phase shift

B. Overload distortion, no phase shift

C. Driving into grid current, and past cut-off, no phase shift

D. Phase shift

E. Phase shift, overload distortion

F. Phase shift, driving into grid current, and past cut-off

G. Amplitude nonlinearity

H. Crossover distortion

(b)

Fig. 4-8. A basic method of distortion analysis. (a) Test setup. (b) Pattern evaluations.

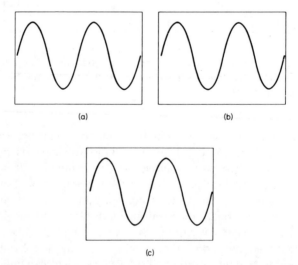

(a)   (b)

(c)

Fig. 4-9. Small percentage distortions are difficult to see in a sine-wave pattern. (a) Negligible distortion. (b) 1 per cent harmonic distortion. (c) 2 per cent harmonic distortion.

It is difficult to evaluate a sine-wave pattern for small percentages of distortion as seen in Fig. 4-9. In other words, a sine wave with 1 or 2 per cent distortion does not appear much different from a sine wave with negligible distortion. However, when Lissajous figures are employed, 1 or 2 per cent distortion is more readily apparent, as shown in Fig. 4-10. To repeat an important point, it is essential to use an oscilloscope that has highly linear vertical and horizontal amplifiers in this type of test work. A laboratory scope is desirable, although there are a few service scopes that have adequately linear vertical and horizontal amplifiers. An oscilloscope can be easily checked for deflection linearity by applying a sine-wave voltage to both the vertical- and horizontal-input terminals. If the resulting Lissajous figure is essentially straight, as in Fig. 4-10(a), the oscilloscope will be satisfactory for checking audio-amplifier linearity.

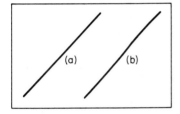 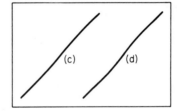

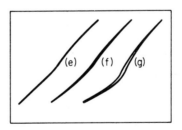

Fig. 4-10. Lissajous patterns with various percentages of distortion. (a) Negligible. (b) 1 per cent harmonic distortion. (c) 1.5 per cent harmonic distortion. (d) 2 per cent harmonic distortion. (e) 3 per cent harmonic distortion. (f) 5 per cent harmonic distortion. (g) 10 per cent harmonic distortion.

Transient analysis is concerned with the response of a circuit or system to a sudden change of input voltage. Square wave tests are commonly employed in transient analysis. A square wave may be regarded as built up from a sine wave and its odd harmonics, as depicted in Fig. 4-11. In theory, a true square wave contains an infinity of harmonics, but in practice, 50 harmonics will provide a reasonable replica of a

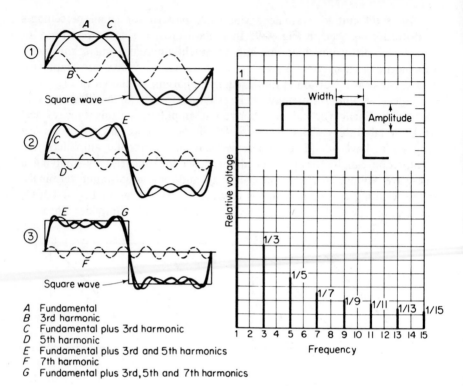

A   Fundamental
B   3rd harmonic
C   Fundamental plus 3rd harmonic
D   5th harmonic
E   Fundamental plus 3rd and 5th harmonics
F   7th harmonic
G   Fundamental plus 3rd, 5th and 7th harmonics

Fig. 4-11. A square wave may be regarded as built up from a sine wave and its odd harmonics.

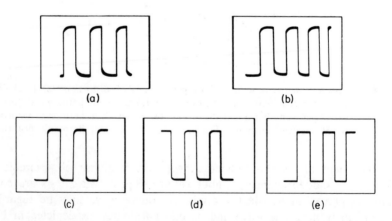

Fig. 4-12. Removal of higher harmonics from a square wave causes corner rounding. (a) 10 harmonics. (b) 25 harmonics. (c) 100 harmonics. (d) 500 harmonics. (e) Over 500 harmonics.

square wave. If a square wave is passed through an amplifier that has insufficient bandwidth to accommodate the necessary harmonics, the reproduced square wave generally becomes rounded at diagonal corners, as shown in Fig. 4-12. Diagonal corners become rounded instead of all four corners because of progressive phase shifts imposed by the amplifier on the harmonics that are passed.

A square-wave test of an audio amplifier is made with the test setup shown in Fig. 4-13(a). It is general practice to use a square-wave repetition rate of 2 kHz, although tests can be made at other repetition rates to obtain additional performance data. Figure 4-13(b) illustrates the specified square-wave response for a particular high-fidelity amplifier. Basic

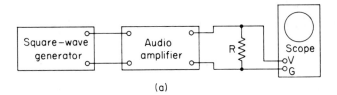

(a)

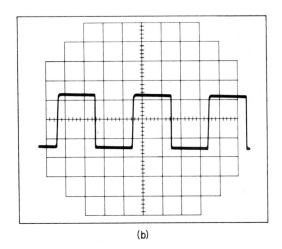

(b)

Fig. 4-13. Square-wave test of audio amplifier. (a) Test setup. (b) Specified 2 kHz square-wave reproduction. (*Courtesy of* General Electric Co.)

types of square-wave distortion are depicted in Fig. 4-14. When an amplifier is being tested, the tone controls should be set to their mid-range positions. That is, low-frequency attenuation or low-frequency boost will occur if the bass tone control is not set to its mid-range position.

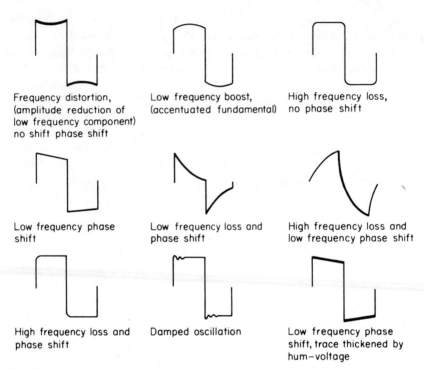

Frequency distortion,
(amplitude reduction of
low frequency component)
no shift phase shift

Low frequency boost,
(accentuated fundamental)

High frequency loss,
no phase shift

Low frequency phase
shift

Low frequency loss and
phase shift

High frequency loss and
low frequency phase shift

High frequency loss and
phase shift

Damped oscillation

Low frequency phase
shift, trace thickened by
hum-voltage

Fig. 4-14. Basic types of square-wave distortion.

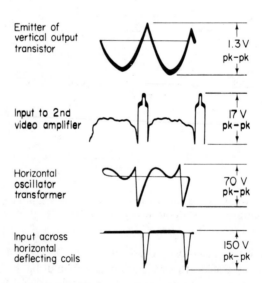

Emitter of
vertical output
transistor

1.3 V
pk-pk

Input to 2nd
video amplifier

17 V
pk-pk

Horizontal
oscillator
transformer

70 V
pk-pk

Input across
horizontal
deflecting coils

150 V
pk-pk

Fig. 4-15. Typical complex waveforms found in TV receivers, with their specified
peak-to-peak voltages.

Similarly, the treble tone control can introduce high-frequency distortion into the reproduced square waveform. It is standard procedure to advance the output from the square-wave generator for maximum rated power output from the amplifier.

Since waveform analysis is an extensive subject, only the most basic principles are noted in this chapter. Both amplitude (peak-to-peak voltage) and waveshape are observed when a complex waveform is being analyzed. Figure 4-15 exemplifies several complex waveforms that occur in television receiver circuitry, with the specified peak-to-peak voltages for each. Note that specified voltages and waveshapes are not absolute. Reasonable tolerances must be taken into account, both on amplitude and waveshape. Thus, an amplitude tolerance of ±20 per cent is considered normal in many circuits. However, some electronic circuits have much tighter tolerances. In any case, the service manual is the authority to be followed. Tolerances on waveshapes involve numerous factors, which are explained in the following chapters.

4.2 BASIC WAVEFORM PROCESSING

Differentiation and integration are related types of basic waveform processing. From a generalized viewpoint, simple filter action is involved in these processes. A complex waveform is differentiated by high-pass filter action, and is integrated by low-pass filter action. RC filter circuits are generally used, as exemplified in Fig. 4-16. When a complex waveform voltage is applied to a series RC circuit, an integrated output waveform appears across the capacitor and a differentiated output waveform appears across the resistor. The sum of the differentiated and integrated waveform amplitudes is equal at any instant to the amplitude of the input waveform, as seen in Fig. 4-16(b). The universal RC time-constant chart shows that the fundamental shape of differentiated and integrated waveforms is always the same, regardless of resistance and capacitance values.

Differentiated and integrated waveforms have an exponential waveshape. This is a mathematical term that describes the equations for the waveforms. As noted previously, the time constant of an RC series circuit is equal to the product of the resistance and capacitance values. In other words, the product of ohms times farads is equal to the time constant in seconds (or fractional second). We observe in Fig. 4-16(c) that differentiated and integrated waveforms have 50 per cent of maximum amplitude at the end of 0.707 time constant. Note that at the end of one time constant, the output from a differentiating circuit has fallen to 37 per cent of its maximum amplitude. Similarly, at the end of one time

93

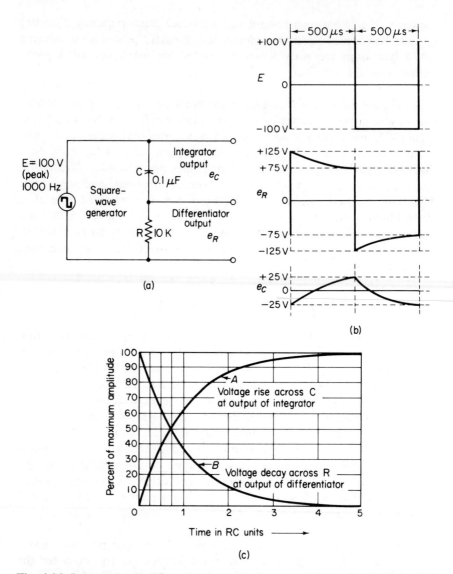

Fig. 4-16. Integrated and differentiated outputs from an RC circuit. (a) Series RC circuit driven by a square-wave generator. (b) Input and output waveforms. (c) Universal RC time-constant chart.

constant, the output from an integrating circuit has risen to 63 per cent of its maximum amplitude. The transient period is practically completed at the end of five time constants.

Exponential waveforms and sine waves are the "building blocks" of

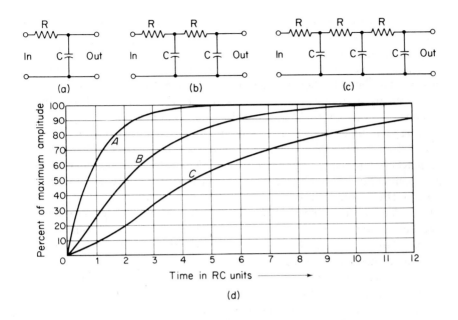

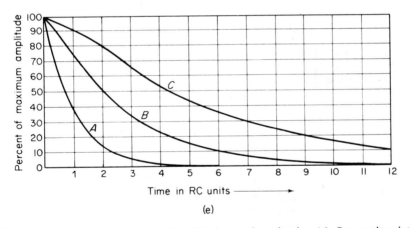

Fig. 4-17. Square-wave responses for RC integrating circuits. (a) One-section integrator. (b) Two-section integrator. (c) Three-section integrator. (d) Leading edges of output waveforms. (e) Trailing edges of output waveforms.

most complex waves. This fact will become evident as we continue our considerations of waveform processing. Figure 4-17 shows square-wave responses for symmetrical two-section and three-section integrating circuits. When an RC section is added to an integrating circuit, the out-

95

put waveforms rise and fall more slowly. Note also that the output waveforms for two-section and three-section integrating circuits are not simple exponential waves. Instead, these waveforms are modified exponential waves; they are particular combinations of simple exponential waves. This modification is caused by the fact that the first RC section is driven by a square wave, whereas the second section is driven by an exponential waveform. Moreover, the first section is loaded by the first section, causing a change in the circuit action of the first section.

Observe next the square-wave response for a symmetrical two-section RC differentiating circuit, shown in Fig. 4-18. The output waveform has a modified or combination exponential waveshape. Comparison with Fig. 4-16(c) shows that the output from a two-section differentiating circuit falls with comparatively great rapidity. Note in Fig. 4-18(b) that the output waveform undershoots the resting level of the circuit to some extent. However, a single-section differentiating circuit does not produce any undershoot. It is instructive in this regard to observe the output from a symmetrical two-section differentiating circuit with sectional isolation, shown in Fig. 4-19. Observe that the undershoot is greater in Fig. 4-19 than it is in Fig. 4-18. The output waveform also falls more slowly when there is sectional isolation. These distinctions result from the fact that

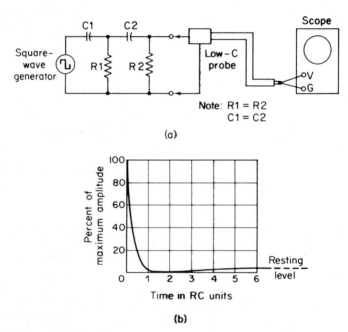

Fig. 4-18. Square-wave response for a symmetrical two-section differentiating circuit. (a) Test setup. (b) Leading edge of output waveform.

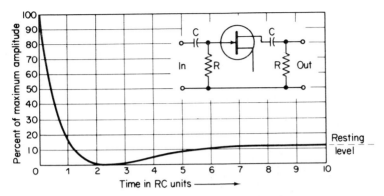

Fig. 4-19. Square-wave response for a symmetrical two-section differentiating circuit with sectional isolation.

the second RC section does not load the first RC section in the configuration of Fig. 4-19.

Next, observe the square-wave response for a symmetrical two-section integrating circuit with sectional isolation (Fig. 4-20). Comparison with Fig. 4-17 shows that isolation between integrating sections causes the output waveform to rise more rapidly. That is, the waveform attains approximately 90 per cent of maximum amplitude in four time constants when sectional isolation is provided, but requires six time constants when the second section is directly connected to the first section. In the configuration of Fig. 4-20, the second section does not load the first section. From another viewpoint, the first section is driven by a square wave, and the second section is driven by a simple exponential waveform.

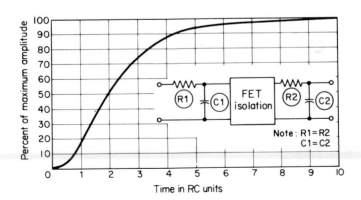

Fig. 4-20. Square-wave response for a symmetrical two-section integrating circuit with sectional isolation.

BASIC OSCILLOSCOPE APPLICATIONS

Transient oscillation (ringing) is another basic type of waveform processing. With reference to Fig. 4-21, a ringing waveform consists of an exponential envelope that encloses a decaying sine wave. Ringing waveforms are produced by shock excitation of LC circuits. A useful application is the measurement of the approximately quality factor Q of a coil. Note that a pulse or square-wave voltage is coupled into the coil

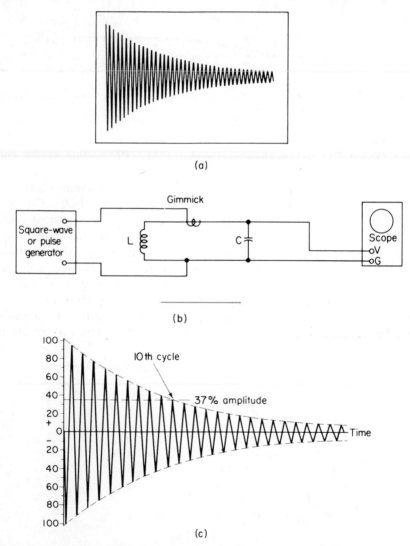

Fig. 4-21. Transient oscillation or ringing waveform. (a) Typical ringing waveform. (b) Test setup for measurement of approximate Q value. (c) Waveform evaluation; Q value is 31.4.

98

by means of a very small capacitor, or with a few turns of wire around the coil lead (gimmick). If the coil normally operates with a shunt capacitor, the capacitor should be included in the test setup. To measure the coil's Q value, observe the 37 per cent level in the waveform and count the number of peaks from the 100 per cent point to the 37 per cent point. This number multiplied by 3.14 gives the approximate Q value. In Fig. 4-21, this value is 31.4.

Note that the ringing frequency can be measured by expanding the waveform horizontally in the manner depicted in Fig. 4-21. Another practical example of a ringing waveform is shown in Fig. 4-22. When the

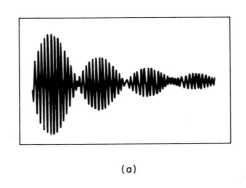

(a)

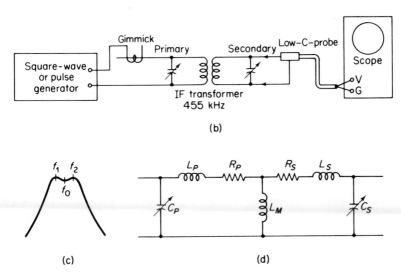

(b)

(c)   (d)

Fig. 4-22. Transient oscillation of a tuned transformer. (a) Typical ringing waveform. (b) Test setup. (c) Double-humped response curve. (d) Equivalent circuit.

primary and secondary of a tuned-IF transformer are resonated at exactly the same frequency, an output ringing waveform is observed as illustrated in (a). Note that the primary and secondary are coupled to give a double-humped frequency-response curve, as shown in (c). The waveform that results is due to the characteristics of the equivalent circuit depicted in (d). Because of the mutual inductance L_M, there are two simultaneous

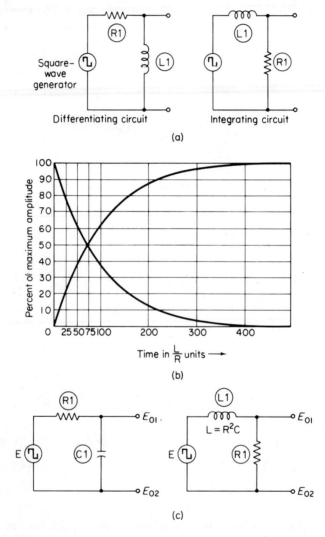

Fig. 4-23. RL differentiating and integrating circuits. (a) Basic circuits. (b) Universal RL time-constant chart. (c) Equivalent RC and RL integrating circuits.

ringing frequencies, f_1 and f_2, where f_0 is the nominal resonant frequency of the transformer.

Although most differentiating and integrating circuits are of the RC type, RL differentiating and integrating circuits are also utilized. Figure 4-23 depicts the basic circuits, and a universal RL time-constant chart. Note that an RC circuit has an equivalent RL circuit. However, as a practical consideration, inductors have appreciable distributed capacitance, which tends to distort the output waveforms from RL circuits. The time constant of an RL circuit is equal to the quotient of the inductance and the resistance. In other words, the inductance in henrys divided by the re-

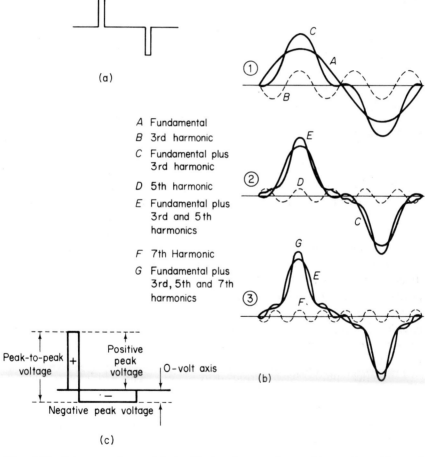

(a)

A Fundamental
B 3rd harmonic
C Fundamental plus
   3rd harmonic
D 5th harmonic
E Fundamental plus
   3rd and 5th
   harmonics
F 7th Harmonic
G Fundamental plus
   3rd, 5th and 7th
   harmonics

(b)

(c)

Fig. 4-24. Pulse waveforms. (a) An ideal pulse waveform with equal positive and negative peaks. (b) Build-up of a pulse waveform from a sine wave and harmonics. (c) An ideal pulse with unequal positive and negative peaks.

101

sistance in ohms is equal to the time constant in seconds (or fraction of a second).

Next, consider the response of a differentiating circuit to a pulse waveform. Figure 4-24 depicts an ideal pulse, and how it can be built up from a fundamental sine wave and its harmonics. Pulse waveforms found in electronic circuitry are less than ideal, and have sloping sides with rounded corners, as exemplified in Fig. 4-25. When a pulse is passed through a

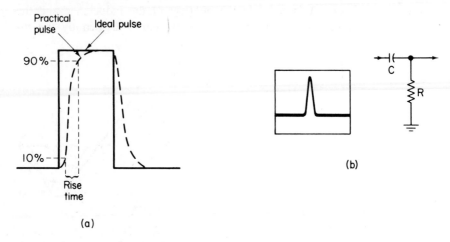

Fig. 4-25. Practical pulse waveform. (a) Example of rise time. (b) Practical pulse and differentiating circuit.

differentiating circuit, the amplitude of the output pulse depends upon the relation of the input pulse rise time t to the time constant T of the differentiating circuit. If T is much greater than t, the output pulse has practically the same amplitude as the input pulse. However, if T is approximately equal to t, the output pulse has approximately half the amplitude of the input pulse. Again, if T is 30 per cent or less of t, the output pulse has approximately 30 per cent or less of the amplitude of the input pulse.

Practical pulses have various waveshapes. The width of a pulse is defined as the elapsed time between the 50 per cent amplitude points on its leading and trailing edges, as shown in Fig. 4-26. Recurrent pulses have a repetition rate, which is related to the elapsed time from the leading edge of one pulse to the leading edge of the next pulse. There is no sharp dividing line between pulses and unsymmetrical square waves. That is, a broad pulse may also be called an unsymmetrical square wave, or a rectangular wave. A pulse is not called a rectangular wave if its width

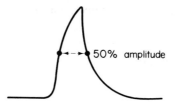

Fig. 4-26. Example of pulse-width measurement.

is small. Note that a square wave is a special case of a rectangular wave, as seen in Fig. 4-27. A pulse may also be of the AC or DC form. An AC pulse has an average value of zero, whereas a DC pulse does not have an average value of zero, owing to a DC component.

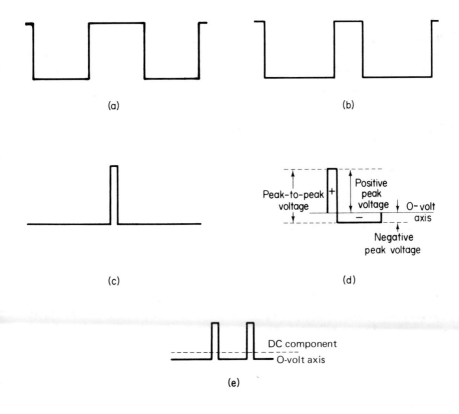

Fig. 4-27. Square, rectangular, and pulse waveforms. (a) Square wave. (b) Rectangular wave. (c) Pulse waveform. (d) AC pulse. (e) DC pulse.

103

Demonstration 1

SUBJECT

RC and RL phase shifting.

OBJECTIVE

To show that:

1. Capacitance causes the current in a circuit to lead the applied voltage.
2. Inductance causes the current in a circuit to lag the applied voltage.
3. The amount of lead or lag depends upon the ratio of L/R or 1/RC, and upon the frequency.

INSTRUCTIONS

This experiment is in the form of a laboratory demonstration, performed by the instructor. A demonstration unit will be utilized to show the effects of capacitance and inductance upon the phase relations between current and applied voltage in a circuit. (See Fig. 4-28.)

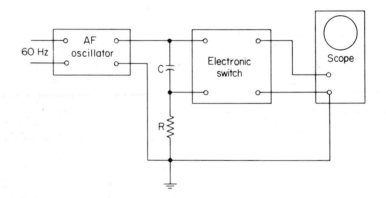

Fig. 4-28. Suggested equipment setup for waveform observation.

Phase shift will be measured visually, by means of an oscilloscope. This measurement is made by noting the horizontal distance between the positive peaks of the voltage and current waveforms. Detailed instructions for this type of measurement are given in a following Information Sheet on phase measurements by the linear sweep method. The student should study and understand this information sheet thoroughly before coming to class. The instructor will insert various combinations of C and R, and of L and R, in the circuit, and will read from the scope screen the amount of phase shift produced by each combination. These figures are to be recorded by the student. Certain calculations are required; these must be performed and recorded before the next laboratory session.

Finally, the instructor will demonstrate the effect of frequency upon the amount of phase shift produced by RC and RL circuits.

PROCEDURE

RC Phase Shifting

1. In Table 4-1, enter the data given by the instructor concerning RC circuits. Leave blank, at this time, the columns reserved for calculated results.

2. Copy the worked-out example of the calculations which will be given by the instructor.

3. After class, perform the indicated calculations, and enter the results in Table 4-1. Attach your work sheets to this table.

TABLE 4-1 Effect of RC Product Upon Phase Shift

R (Ohms)	C (μf)	X_c (Ohms)	X_c/R	CALCULATED PHASE SHIFT	MEASURED PHASE SHIFT
470,000	.1				
47,000	.1				
33,000	.1				
22,000	.1				
10,000	.1				
5,000	.1				
470	.1				

RL Phase Shifting

1. In Table 4-2, enter the data given by the instructor concerning RL circuits. Leave blank, at this time, the columns reserved for calculated results.

2. Copy the worked-out example that will be given by the instructor.

3. After class, perform the indicated calculations, and enter the results in Table 4-2. Attach your work sheets to this table.

TABLE 4-2 Effect of RL Ratio Upon Phase Shift

R (OHMS)	L (HENRIES)	X_L (OHMS)	X_L/R	CALCULATED PHASE SHIFT	MEASURED PHASE SHIFT
470,000					
47,000					
33,000					
22,000					
10,000					
5,000					
470					

Note: The instructor will furnish all necessary information concerning the true inductance and effective resistance of the choke coil used as L.

Effect of Frequency

1. Copy any rules and examples of the effect upon phase shifting that the instructor may give you.
2. Work out any problems assigned by the instructor. Show all your calculations.

> **Note:** An electronic switch is an accessory unit that converts a single-trace oscilloscope into a dual-trace oscilloscope. If a dual-trace oscilloscope is available, the electronic switch is not used in the demonstration equipment setup.

CONCLUSIONS

Summarize, in not less than 100 words, the facts brought out in this demonstration concerning:

1. The amount and direction of phase shift introduced by the presence of RC and RL combinations in a circuit.
2. The effects of frequency upon phase shift.
3. The relationship between phase shift and time delay.
4. Some possible applications of phase-shifting networks in electronic procedures.

Information Sheet—
Phase-Shift Measurement
by the Linear-Sweep Method

It is often desirable or necessary to know the precise phase relationship between two voltages, two currents, or a voltage and a current in circuits that are being tested. A moderately accurate method of measuring phase difference with an oscilloscope and electronic switch is described below.

EQUIPMENT REQUIRED

Equipment for this measurement is set up as shown in Fig. 4-29. Channel *A* of the electronic switch is connected across the input voltage impressed upon the phase-shifting network RC. This channel passes on to the oscilloscope a waveform that represents accurately the shape and phase of the input voltage *E*. Channel *B* of the electronic switch is connected across the resistor R of the phase-shifting network. Since this network is a series circuit, all of the current in the circuit flows through R, producing a voltage drop across it. This voltage drop is exactly in phase with the current flow, and can be used to display a voltage waveform on the CRT screen which has the same waveform and the same phase as the current waveform.

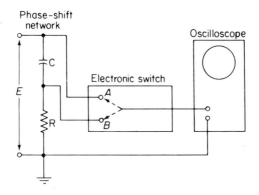

Fig. 4-29. Equipment used for measuring phase shift.

The two channels of the electronic switch are switched on and off alternately at a very rapid rate. Only one channel passes voltage to the scope at any one instant. This switching action takes place so rapidly that the two traces—one produced by the output of channel *A,* and the other produced by the output of channel *B*—appear simultaneously on the CRT screen in the form of dotted lines.

107

Phase-Difference Measurement

The procedure for making a measurement of the phase difference between input voltage *E* and the circuit current *I* is as follows:

1. Adjust the oscilloscope controls to display three cycles of the

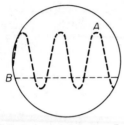

(a) Adjust channel *A* to show 3
    cycles
    Adjust channel *B* to zero

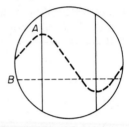

(b) Spread channel *A* until its peaks
    are 10 divisions apart; channel
    B still at zero

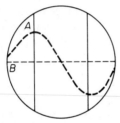

(c) Adjust trace—seperation control
    until channel *B* trace bisects
    channel *A* trace

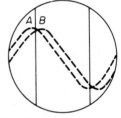

(d) Adjust channel *B* trace to same
    size as channel *A* trace

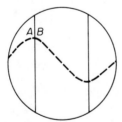

(e) If the traces coincide exactly,
    the two waves are in phase

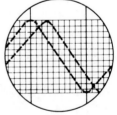

(f) If the two traces do not exactly
    coincide, count the number of
    squares between their
    corresponding peaks, and
    multiply by 18

Fig. 4-30. Phase-difference measurement.

108

waveforms. Turn the gain control of channel *B* to zero. This trace will then become a straight line, drawn across the channel *A* waveform, as depicted in Fig. 4-30(a).

2. Expand the scope trace horizontally until the positive peak of the middle cycle rests upon one of the heavy vertical black lines on the graticule, while the negative peak rests upon the next heavy vertical line to the right (ten small squares away), as depicted in Fig. 4-30(b).

   (a) Use of the middle cycle of a group of three minimizes distortion which might occur in either of the end cycles, due to residual end effects in time-base operation.

   (b) Since the two peaks are 180 deg apart, and since they have been adjusted so that they are separated by 10 squares, each horizontal square represents 18 deg.

3. Adjust the trace-separation control on the electronic switch until the dotted straight line mentioned in step 1 falls exactly at the center of the sine-wave trace from the other channel, as depicted in Fig. 4-30(c).

4. Now advance the gain control of the channel that has been turned off until its trace is approximately the same height as that produced by the other channel, as depicted in Fig. 4-30(d).

5. If the two traces can be made to coincide exactly, and to form a single sine wave, it indicates that the current and applied voltage are exactly in phase. [See Fig. 4-30(e).]

6. If, however, the two sine waves do not coincide, this indicates that a phase difference exists between the voltages represented by the two traces. This difference, in degrees, can be determined by counting the number of horizontal squares between the positive peak of the channel *A* trace, and the positive peak of the channel *B* trace. As an illustration, if these peaks are separated by three squares, each square represents 18 deg, and in turn the voltage and current are 54 deg out of phase with each other. [See Fig. 4-30(f).]

DISTINGUISHING LEAD FROM LAG

Having determined the number of degrees of phase difference between the two waveforms, it is now necessary to determine whether the current (channel *B*)leads or lags the applied voltage (channel *A*). Refer to Fig. 4-31 and proceed as follows:

1. Select one of the traces for use as a reference. (In this example we have selected the applied voltage, channel *A*.)

109

2. Locate the point at which this reference waveform is passing through zero and preparing to enter its positive alternation. (This point is indicated in Fig. 4-31.)

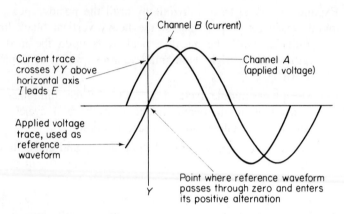

Fig. 4-31. Method of determining whether the current (channel B) leads or lags the reference voltage (channel A).

3. Draw a vertical line through this point. (This is the vertical line YY in Fig. 4-31.)

4. Observe whether the trace whose phase is being compared with the reference waveform crosses this vertical line above or below the horizontal axis. (In this example, the trace from channel B crosses the vertical line YY above the horizontal axis.)

5. If it crosses above the horizontal axis, the waveform being compared leads the reference waveform. Since the channel B trace (circuit current) crosses the vertical line above the horizontal axis, we know that the current in the circuit leads the applied voltage (channel A). This is in accordance with theory which states that, in a capacitive circuit, current leads the applied voltage. The amount of lead is 54 deg, as we already determined in a previous step.

6. If the trace crosses YY below the horizontal axis, the waveform being compared lags the reference waveform.

Demonstration 2
Phase-Shift Measurement

In the previous demonstration, an electronic switch was employed to produce dual traces on the CRT for measurement of phase shift. Now, we

110

shall note how to measure phase shift by using successive traces and the external-sync function of the oscilloscope. With reference to Fig. 4-32, a capacitor and a resistor are connected in series, and oscilloscope connections are made to measure the phase difference between the circuit current and the applied voltage. Note that the applied voltage (secondary of the step-down transformer) is connected to the external-sync terminal of the oscilloscope. In turn, the horizontal sweep will always start in time with the applied voltage, regardless of the signal that is applied to the vertical-input terminal of the oscilloscope.

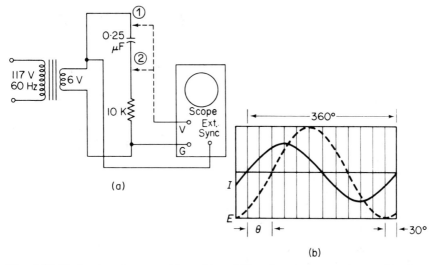

Fig. 4-32. Test setup for measuring phase shift using the external-sync function. (a) Connections. (b) CRT patterns.

The first check is made at point 1, which applies the source voltage to the vertical-input terminal. It is convenient to adjust the horizontal-amplifier control so that one complete cycle occupies 12 divisions. In turn, each division of the graticule represents 30 deg. With the pattern centered on the screen, note the point at which the voltage waveform goes through zero. Then shift the vertical-input lead of the oscilloscope to point 2. In turn, the current waveform is displayed on the screen, because the voltage drop across a resistor is in phase with the current flowing through the resistor. Note that point at which the current waveform goes through zero, and count divisions between this point and the point at which the voltage waveform went through zero. As an illustration, if there should be two divisions between the current and voltage zero points, the phase difference is 60 deg.

111

5

SIGNAL TRACING
AND
TROUBLESHOOTING
PROCEDURES

5.1 GENERAL CONSIDERATIONS

Although there is no sharp dividing line between signal-tracing and trouble-shooting procedures, the following distinctions may be noted:

1. Signal tracing analyses involve a signal flow-chart viewpoint, and are chiefly concerned with the presence or absence of signal voltages at progressive key test points through an electronic unit or system.

2. Troubleshooting analyses involve evaluation of waveform characteristics, with the purpose of determining the cause of a malfunction, the nature of the fault, and sometimes the identification of a defective component.

As an illustration, Fig. 5-1 depicts the key signal-tracing points in a simple audio amplifier. In one basic aspect, a block diagram is a signal-flow chart. If, for example, we find a sine-wave pattern at test-point 1, but little or no pattern at test-point 2, we shall conclude that the trouble will

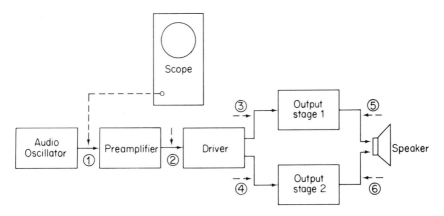

Fig. 5-1. Key signal-tracing points in a simple audio amplifier.

be found in the preamplifier stage. Again, if we find a sine-wave pattern at test-point 2, but little or no pattern at test-points 3 and 4, we shall conclude that the trouble will be found in the driver stage. Thus, the foregoing signal-tracing tests serve to localize an operating fault to a particular section of the amplifier system. On the other hand, the signal-tracing tests do not indicate what component or malfunction in the section is causing the trouble symptom.

Next, consider a trouble situation in which the sound output from an audio amplifier is distorted. In such a case, it is desirable to use Lissajous patterns, because they are easier to evaluate for distortion than are sine-wave patterns. As an illustration, it has been mentioned previously that it is very difficult to observe 1 or 2 per cent harmonic distortion in a sine wave, whereas this amount of distortion can be definitely perceived by examination of a Lissajous pattern. Note that Lissajous patterns can be used to analyze individual stages, or to analyze system operation. For example, Fig. 5-2 shows how a preamplifier is checked for distortion, and Fig. 5-3 shows how a complete audio amplifier is checked. There are some essential points to be observed in this type of test work:

1. The oscilloscope amplifiers must be rated for a lower percentage of distortion than the audio amplifier or system under test.

2. The audio amplifier under test should be driven at full rated power output on all tests except for crossover distortion. (Crossover distortion becomes more prominent at low operating levels).

3. If the speaker is disconnected from the output stage(s), an equivalent power resistor should be connected in its place to provide a normal output load.

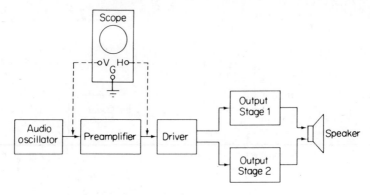

Fig. 5-2. Testing the preamplifier stage for distortion.

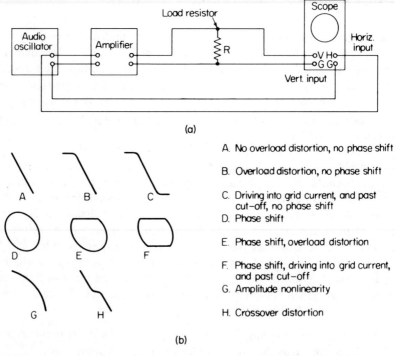

A. No overload distortion, no phase shift

B. Overload distortion, no phase shift

C. Driving into grid current, and past cut-off, no phase shift

D. Phase shift

E. Phase shift, overload distortion

F. Phase shift, driving into grid current, and past cut-off

G. Amplitude nonlinearity

H. Crossover distortion

(b)

Fig. 5-3. Audio amplifier operating tests. (a) Test connections. (b) Pattern evaluations.

In normal operation at a 1-kHz test frequency, the Lissajous pattern will be a straight diagonal line, as depicted in Fig. 5-3(a). However, as the test frequency is increased toward the high-frequency cutoff point of the amplifier, phase shift will become evident, and the diagonal line will open up into a diagonal ellipse. Similarly, phase shift becomes evident as the test frequency is decreased toward the low-frequency cutoff point of the amplifier. (Note that a DC-coupled amplifier does not have a low-frequency cutoff point.)

Note that when audio-amplifier distortion is evaluated on the basis of sine-wave patterns, the audio-oscillator output must provide an output waveform that has a lower percentage of distortion than the audio amplifier under test. Otherwise, any distortion contributed by the audio oscillator will be falsely charged to the amplifier under test. On the other hand, when audio-amplifier distortion is evaluated on the basis of Lissajous patterns, the demand on audio-oscillator waveform purity is greatly relaxed. In other words, an audio oscillator might have several percent harmonic distortion in its output waveform, but a perfectly straight diagonal line will still be displayed on the CRT screen, if the audio amplifier under test is distortionless.

5.2 OSCILLOSCOPE TESTS WITH PUSH-PULL INPUT

Most oscilloscopes have single-ended input. This means that one of the input terminals is "hot," whereas the other input terminal is grounded to the case of the oscilloscope. When push-pull amplifiers are to be tested, it is sometimes helpful to employ an oscilloscope that has push-pull input (double-ended input). Figure 5-4 shows the plan of an oscilloscope with double-ended vertical input. Note that three input terminals are provided, identified as Vert Input No. 1, Gnd, and Vert Input No. 2. Each vertical-input signal drives a transistor, and both transistors have a common source resistor R_s. In turn, the amplified outputs from Q1 and Q2 are applied to the vertical-deflecting plates of the CRT.

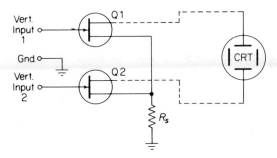

Fig. 5-4. Plan of an oscilloscope with double-ended vertical input.

This double-ended input arrangement can be used to test either single-ended or push-pull circuits. For example, to operate the input circuit as a conventional single-ended oscilloscope, a shorting link can be connected between Vert Input No. 2 and Gnd. In turn, a signal voltage from a single-ended circuit applied between Vert Input No. 1 and Gnd is displayed on the CRT screen in the usual manner. Note that the CRT is being driven in push-pull, because a signal voltage applied to the gate of Q1 causes a signal-voltage drop across R_s. In turn, Q2 is source-driven by R_s, and an inverted signal appears at the drain of Q2. Thus, Q2 operates as a simple phase inverter in this situation.

Next, let us consider a distortion check of the driver stage in Fig. 5-2. This driver stage has single-ended input and double-ended output. We can use a single-ended oscilloscope for this test, and check each output separately (make two separate tests). As long as the driver operates in class *A*, the test results are clear. On the other hand, if the driver operates in class *AB*, it becomes difficult to combine the results of the individual tests and arrive at a very accurate conclusion concerning the overall linearity of stage operation. Therefore, it is helpful to use a double-ended oscilloscope in the test setup shown in Fig. 5-5. A single test is made, and because the oscilloscope action is such as to combine the two outputs in a resultant pattern, an accurate conclusion concerning overall linearity of stage operation can be arrived at immediately.

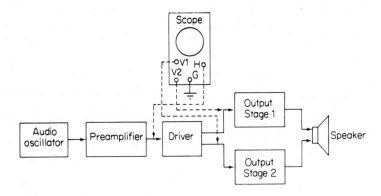

Fig. 5-5. Linearity check of a driver stage with single-ended input and double-ended output.

As shown in Fig. 5-6, audio oscillators are also available that provide a choice of single-ended or double-ended output. When single-ended output is utilized, a shorting link is connected between one "hot" terminal and ground, as from 1 to *G*. The output signal is then available between ter-

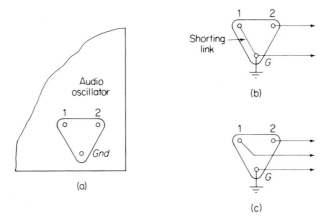

Fig. 5-6. Audio-oscillator terminal board that provides either single-ended or double-ended output.

minals 2 and *G*. On the other hand, when double-ended output is utilized, the shorting link is omitted. The push-pull output signal is then provided by terminals 1 and 2, with respect to the ground terminal *G*. It is often helpful to have a push-pull test signal available when one is checking output amplifiers that operate with push-pull input. As a practical trouble-shooting note, service data for audio amplifiers may not specify pertinent test-signal levels. In such a case, the technician must fall back upon his experience and his ability to analyze amplifier action. Sometimes, a comparison signal-tracing check can be made on a similar amplifier that is in good condition.

5.3 SIGNAL TRACING AND TROUBLE-SHOOTING TAPE-RECORDER CIRCUITRY

As seen in Fig. 5-7, a simple tape recorder includes a preamplifier, amplifier, driver, and push-pull output stage. In addition to the amplifier channel, a bias oscillator is provided. To signal-trace the amplifier channel, the system may be energized from the microphone-input jack, the auxiliary high-level input jack, or the playback head. Note that the output level from a playback head is typically 8 millivolts. Few audio oscillators can supply a low-level test signal. Therefore, it is usually necessary to utilize an external attenuator arrangement if an audio oscillator is used as a signal source. As an illustration, an audio oscillator might have a minimum signal output

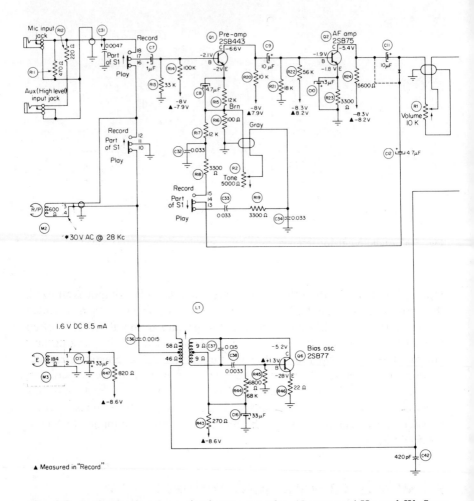

Fig. 5-7. Audio circuitry for a simple tape recorder. (*Courtesy of* Howard W. Sams & Co., Inc.)

level of 0.1 volt. In turn, a voltage divider as depicted in Fig. 5-8 can be employed to obtain an 8-millivolt test signal.

An audio-oscillator signal might be applied, for example, at either the left-hand end or the right-hand end of C7 in Fig. 5-7. If the signal is applied at the left-hand end of the capacitor, the output from the divider depicted in Fig. 5-8 can be used directly. On the other hand, if the signal is applied at the right-hand end of the capacitor, it is essential to insert a blocking capacitor in series with the "hot" lead from the voltage divider. Otherwise, the −2.1 volt bias at the base of Q1 will be reduced to zero. In other

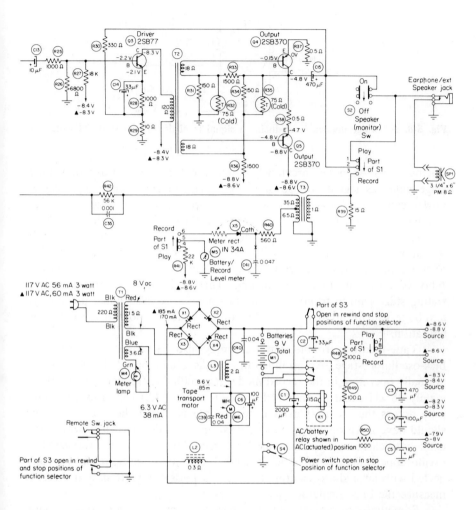

words, the 1-ohm resistor in the divider circuit of Fig. 5-8 will drain off the DC bias voltage from the transistor. Although the value of blocking capacitor chosen in this application is not critical, a 1-μf capacitor is appropriate, since this is the value of the coupling capacitor used in the amplifier circuit. If a blocking capacitor were not used in this situation, the emitter of Q1 would be biased -2 volts with respect to the base, and Q1 would be cut off. In turn, the technician would falsely conclude that the preamplifier stage was "dead."

Sometimes gain measurements are made in combination with signal-

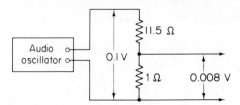

Fig. 5-8. A voltage divider for reducing a signal level of 0.1 volt to 0.008 volt.

tracing tests. As noted previously, there are three basic ways of stating stage gain: as signal voltage gain, as signal current gain, and as signal power gain. The voltage gain of the preamplifier stage in Fig. 5-7 is determined by noting the vertical deflection on the CRT screen when checking the signal at the base of Q1, and comparing this value of deflection with the amount of deflection obtained at the collector of Q1. It is good practice to use a low-capacitance probe with the oscilloscope, to avoid the possibility of error due to circuit loading. This is the most common way of stating stage gain in troubleshooting procedures. As also noted previously, a voltage gain of approximately 250 times would be normal for this configuration.

Tape-recorder trouble symptoms may also be caused by a defective bias-oscillator stage. As as illustration, bias oscillator Q6 in Fig. 5-7 is normally operating in the record mode. Its oscillating frequency is approximately 28 kHz, and its amplitude is approximately 30 volts rms (85 volts p-p) across the record/playback head. Note also that the bias waveform should have a reasonably good sine waveshape. A distorted bias waveform is likely to cause background "hiss" in a recording. Although a service-type oscilloscope cannot be used to measure frequency directly, it can be connected with an audio oscillator to display a Lissajous pattern, and thereby measure the bias-oscillator frequency indirectly.

Sometimes frequency-response checks are made in combination with signal-tracing tests. As an illustration, an 8-millivolt signal from an audio oscillator may be applied across the record/playback head M2 in Fig. 5-7, and the oscilloscope (with a low-capacitance probe) connected at the collector of Q3. In turn, the operating frequency of the audio oscillator can be varied from 20 Hz to 15 kHz, and the CRT pattern observed for any serious change in amplitude. Of course, it is necessary to use an audio oscillator that has a uniform output amplitude over this frequency range, or the variation would be falsely charged to the amplifier stages under test. Note that the tone control must be set to the midpoint of its range, or the frequency response will appear to be nonuniform. The general effect of treble and bass boosting and cutting is seen in Fig. 5-9.

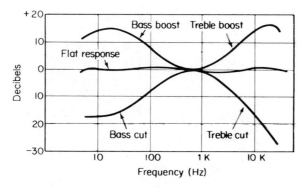

Fig. 5-9. Response curve for bass and treble tone controls.

Beginners sometimes overlook the fact that equalizers are used in tape-recorder amplifiers. In turn, the normal frequency response of the amplifier system is not uniform but follows a playback equalization curve, as shown in Fig. 5-10. Since few oscilloscopes are provided with dB graticules, it is necessary to convert the peak-to-peak voltage readings into

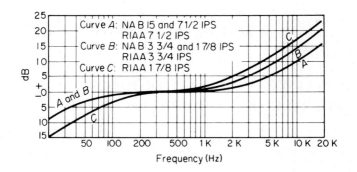

Fig. 5-10. Standard playback equalization curves in use for various tape speeds.

corresponding dB values. This is easily done by reference to Fig. 5-11, which shows the relation of voltage ratios from 1 to 1000 versus decibel values from zero to 60. It is instructive to note that the equalization network in Fig. 5-7 is in the emitter circuit of Q1. The playback equalization curve is affected in part by the feedback loop from Q2 to Q1 via C12.

121

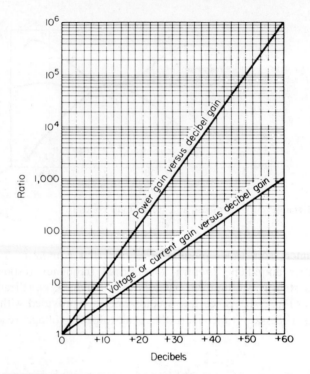

Fig. 5-11. Gain ratio plotted against decibel gain. This chart may also be used for attenuation by inverting the ratio and changing the dB sign from positive to negative.

5.4 SIGNAL TRACING AND TROUBLE-SHOOTING OF AM RADIO RECEIVERS

Signal tracing of an AM radio receiver can be accomplished to best advantage with the aid of an AM signal generator. The generator supplies a steady signal which can be set to any desired frequency, amplitude, and percentage of modulation by the technician. Preliminary signal-tracing procedures are usually made with a weak output signal from the generator, so that the AVC section of the receiver is "wide open." Subsequent tests can sometimes be made to better advantage with the AVC section clamped at some particular voltage. The oscilloscope must be used with a low-capacitance probe. If a 100-to-1 low-capacitance probe is employed, circuit loading will be minimized in the RF, oscillator, mixer, and IF stages. In turn, the test conclusions will be more reliable.

Figure 5-12 shows a test setup for signal-tracing an AM radio receiver. Key checkpoints are numbered from 1 to 10. Note that if the input

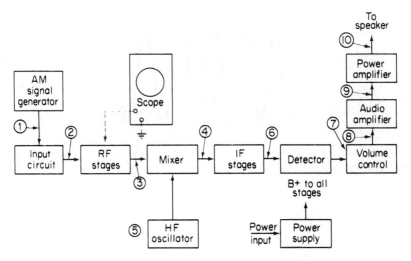

Fig. 5-12. Test setup for signal-tracing an AM radio receiver.

circuit of the receiver is tuned to 1 MHz, for example, the signal generator must also be tuned to 1 MHz. The RF signal is usually modulated at either 400 Hz or at 1 kHz, and at a percentage modulation from 30 to 100 per cent. When 100 per cent modulation is employed, the scope display will appear as illustrated in Fig. 5-13. This waveform is normally found at test points 1 through 6, with one exception. That is, the waveform at test point

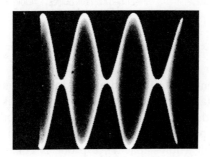

Fig. 5-13. AM test signal with 100 per cent modulation.

5 is normally a sine wave, as illustrated in Fig. 5-14. Waveform changes that normally occur through the detector section are depicted in Fig. 5-15.

Typical signal-voltage gains for each stage of a small AM radio receiver are noted in Fig. 5-16. To obtain reliable measurements, the gen-

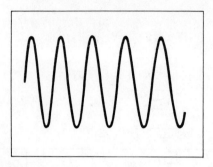

Fig. 5-14. Normal oscillator output waveform.

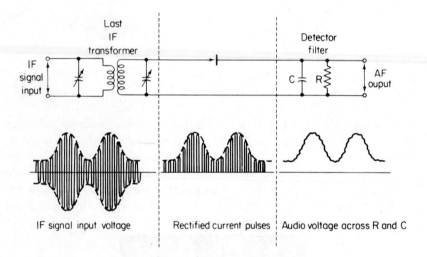

Fig. 5-15. Waveform changes that occur through the detector section.

erator should be connected (through a blocking capacitor of about 0.1 μf) at the input of the stage under test. A comparatively weak output signal level should be employed, to avoid activating the AVC section. The generator must be set to the pertinent operating frequency, such as 600 kHz. Then the oscilloscope with a low-capacitance probe is connected in turn at the input and at the output of the stage. The ratio of the two vertical-deflection amplitudes is then equal to the stage gain. The chief requirement for the scope is high vertical sensitivity, so that ample vertical deflection can be obtained. A bandwidth of 2 MHz is sufficient for signal tracing conventional AM receivers. High sensitivity is often needed, such as 10 millivolts per inch.

Note that the output from a mixer or converter stage contains a mix-

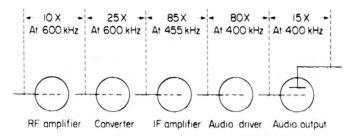

Fig. 5-16. Typical signal-voltage gains for each stage of a small AM radio receiver.

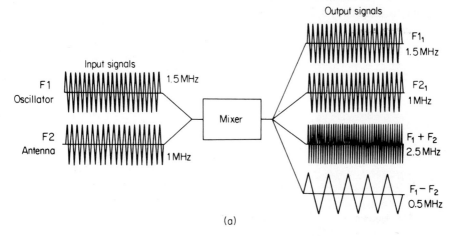

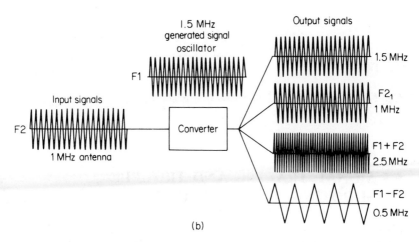

Fig. 5-17. Input and ouptut signal frequencies for a mixer stage and a converter stage. (a) Mixer. (b) Converter.

125

ture of frequencies, as exemplified in Fig. 5-17. This mixture consists of the applied test frequency, the oscillator frequency, the sum of these two frequencies, and the difference of these two frequencies. Since tuned circuitry is utilized in this stage, the dominant frequency displayed on the CRT screen depends upon the particular tuned circuit across which the oscilloscope is applied. As an illustration, with reference to Fig. 5-18, frequency F2 is very strong at the base of Q1. On the other hand, F1-F2 is very strong at the collector of Q1. In other words, T3 is tuned to frequency F1-F2, and largely rejects the other component frequencies. Frequency F1 dominates the waveform at the emitter of Q1 because T2 is tuned to frequency F1.

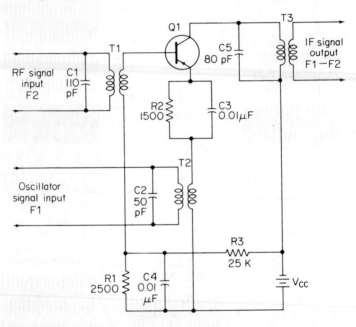

Fig. 5-18. Distribution of frequencies in a mixer circuit.

5.5 SIGNAL TRACING AND TROUBLE-
SHOOTING OF FM RADIO RECEIVERS

Most of the basic principles that are involved in signal tracing AM radio receivers apply also to signal tracing FM receivers. However, there are some practical differences which must be observed. With reference to Fig.

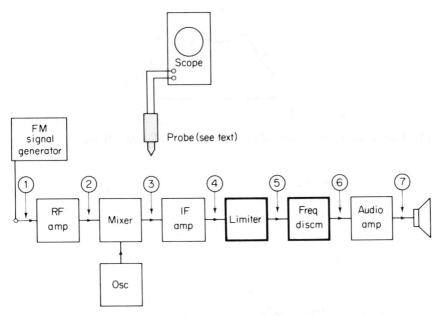

Fig. 5-19. Test setup for signal tracing an FM radio receiver.

5-19, it is desirable to use an FM generator as a signal source. Inasmuch as FM demodulator probes are not available for signal tracing, the test signal must be effectively amplitude-modulated, as well as frequency-modulated. This is easily accomplished by employing an FM sweep generator set for maximum sweep width. In turn, the tuned signal channel develops a frequency-response curve that can be displayed on the CRT screen for tracing the signal.

Since the FM frequency band extends from 88 to 108 MHz, a service-type oscilloscope cannot respond directly to a test signal in this range. Therefore, a demodulator probe must be used with the oscilloscope to check the signal through the RF section. The IF operating frequency is 10.7 MHz, and this is also beyond the response capability of the great majority of service-type oscilloscopes. Accordingly, a demodulator probe must also be used to trace the signal through the IF section. Figure 5-20 illustrates a typical CRT pattern obtained in this method of signal tracing. Since a demodulator probe loads the tuned circuits under test rather heavily, stage-gain measurements are unreliable with this method. Its chief utility is to show whether the test signal is present or absent at successive stages.

The limiter stage is checked by applying the demodulator probe at the limiter output and varying the output level from the FM sweep gen-

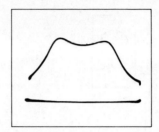

Fig. 5-20. Typical CRT pattern obtained in signal tracing an FM receiver.

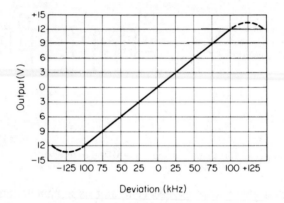

Deviation (kHz)

Fig. 5-21. Normal response for an FM discriminator.

erator through a substantial range. If the limiter is operating properly, the vertical height of the CRT pattern remains practically constant as the signal level is varied. Note that the shape of the pattern is inconsequential. Next, the discriminator stage is checked by using a low-capacitance probe with the oscilloscope and observing the waveform at the discriminator output. In normal operation, this is an S curve, as illustrated in Fig. 5-21. Similarly, the S-curve pattern is normally observed at the output of the audio amplifier.

5.6 SIGNAL TRACING IN STEREO-MULTIPLEX CIRCUITRY

Stereo-multiplex circuitry is a part of a sound system, as exemplified in Fig. 5-22. The stereo-multiplex circuitry is enclosed within the dotted lines. Note that the stereo-multiplex input signal is obtained from the ratio detector, and the output signals are applied to a pair of audio-amplifier

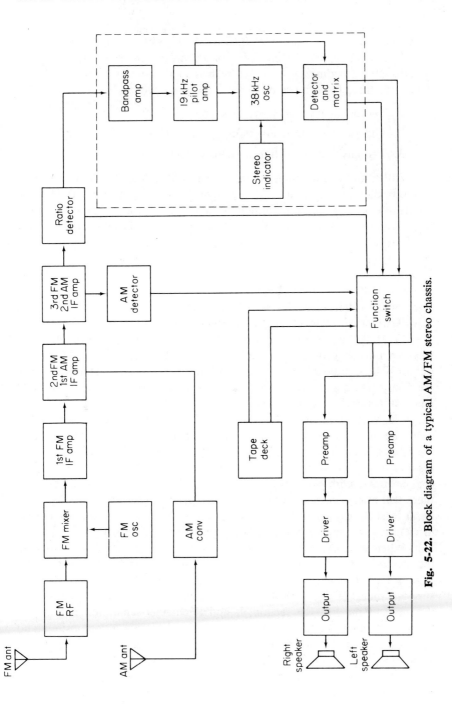

Fig. 5-22. Block diagram of a typical AM/FM stereo chassis.

channels. Thus, signal tracing in stereo-multiplex circuitry involves the procedures that have been explained for FM-radio signal tracing, and for audio-amplifier signal tracing, plus procedures that are pertinent to the stereo-signal decoding process that occurs in the stereo-multiplex section. There are three basic types of stereo-multiplex decoders:

1. The bandpass and matrix system.

2. The switching-bridge system.

3. The envelope-detection system.

Although the end result is the same in each case, the signal-processing methods differ more or less. Consequently input-signal and output-signal checks are always the same, although the waveforms and signal flow paths will differ within these basic types of decoders. The basic function of the decoder, as follows from the block diagram of Fig. 5-22, is to process the encoded signal from the ratio detector so that its inherent left- and right-channel components are separated for application to the left and right audio channels. Effective signal tracing requires the use of a stereo-multiplex generator, so that a steady test signal is available which can be switched at will from left-channel output to right-channel output.

Figure 5-23 shows the complete composite audio signals provided by a stereo-multiplex generator for left-channel output and for right-channel output. These waveforms appear to be the same at first glance, because their only difference is in the phasing of the waveform components. L and R channel separation is the most basic test, as depicted in Fig. 5-24. For example, with an R-channel signal applied from the generator, an oscilloscope is utilized to observe the relative amplitudes of signal outputs from the L and R audio channels. Figure 5-25 illustrates ideal and typical test results. In the ideal situation, there would be an output signal from the R channel only, with no output from the L channel. However, in practice there will always be some output from the L channel. Normally, there will be at least 25 dB separation. In the example of Fig. 5-25, the waveforms in (a) and (c) show a separation of approximately 18 dB.

When unsatisfactory separation occurs, the trouble may be found in the stereo-multiplex section, or it may be found in the FM-radio section. Therefore, localization tests are required. With reference to Fig. 5-22, suppose that when a 100-MHz stereo-multiplex test signal is applied at the input of the FM RF section, L and R channel separation is found to be unsatisfactory. In such a case, we would then apply a composite audio stereo-multiplex test signal at the input of the bandpass amplifier. If the separation value is then satisfactory, the trouble will be found in the FM-radio section (possibly the tuned circuits are in need of alignment). Otherwise, there is a defective component in the stereo-multiplex section.

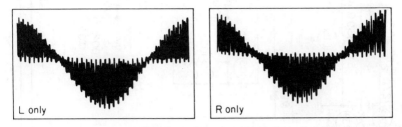

Fig. 5-23. Complete composite audio signals provided by a stereo-multiplex generator.

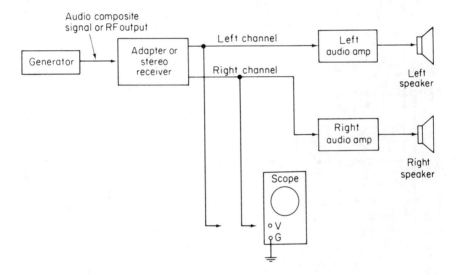

Fig. 5-24. Basic stereo-multiplex separation test.

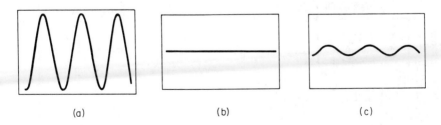

Fig. 5-25. Waveforms observed during separation test. (a) R channel output. (b) Ideal L channel output. (c) Typical L channel output.

131

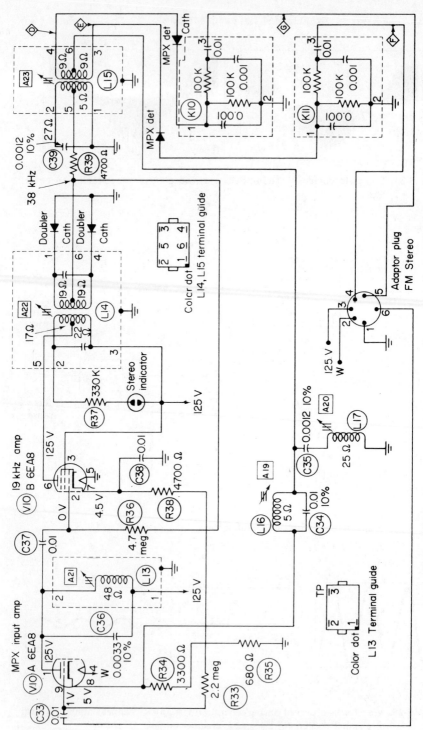

Fig. 5-26. Configuration of an envelope-detector type of stereo-multiplex decoder.

132

Figure 5-26 exemplifies the envelope-detection type of decoder. Assume that the R Only test signal in Fig. 5-23 is being applied. This signal has two chief components: the encoded audio information and a 19-kHz pilot subcarrier. The decoder processes the composite stereo-multiplex signal by trapping out the 19-kHz pilot subcarrier, and passing it through a doubler. In turn, the 38-kHz subcarrier is recovered. (The subcarrier is not transmitted, being suppressed at the transmitting station.) When the encoded audio information is combined with the recovered 38-kHz subcarrier in the secondary of L15, a waveform as shown in Fig. 5-27(a) normally appears at point D. A similar waveform normally appears at point E. Since the multiplex detector diodes are oppositely polarized, it follows that the R waveform will be displayed normally at point G, whereas the L waveform will be displayed normally at point F.

With reference to Fig. 5-27(b), the L waveform could be an audio signal that differs from the R waveform in both frequency and amplitude. If this undistorted waveform were applied to the multiplex detector diodes, separation would be ideal. On the other hand, no multiplex decoder provides ideal operation, and 25 or 30 dB separation represents typical normal operation. Poor separation is most likely to be caused by defective capacitors, or by diodes with impaired front-to-back ratios. Capacitor leakage often causes disturbance of DC bias values, with the result that multiplex decoder amplifiers operate nonlinearly. It follows from Fig. 5-27 that if a multiplex detector diode has a poor front-to-back ratio, some of the R envelope signal will cross over into the L envelope region, and vice versa.

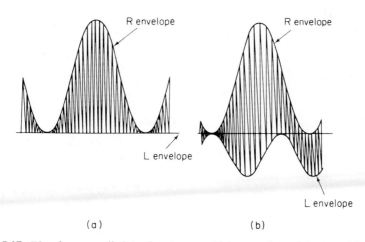

(a)   (b)

Fig. 5-27. Waveforms applied to the stereo-multiplex envelope detectors. (a) Right-channel test signal only. (b) Right- and left-channel test signals with different audio frequencies.

5.7 SIGNAL TRACING IN INDUSTRIAL ELECTRONICS CIRCUITRY

Many types of circuits and systems are used in industrial electronics equipment. Power sources are basic, as in other areas of electronics. However, variable power sources that may supply high current demands are often encountered in industrial electronics systems. As a comparatively simple example, Fig. 5-28 shows a thyratron circuit that supplies a variable DC current I through the load in response to the setting of a DC control. The thyratron tubes V1 and V2 pass a value of current that depends upon their conduction angle. In other words, a thyratron does not rectify over a complete half cycle, except as a special case (maximum conduction). If the grid voltage of a thyratron is 180 deg out of phase with its anode voltage, there is zero current flow from cathode to anode. As the phase of the grid voltage is shifted, the current flow changes accordingly.

It is helpful to understand the circuit action that occurs in Fig. 5-28. The source of the grid voltage is transformer T2. Note that T1 is the anode or main power-supply transformer. The relative phase of the thyratron grid voltage is adjusted by a phase-shifting network, called a dephasing circuit. This dephasing circuit comprises T3, R, the saturable-core

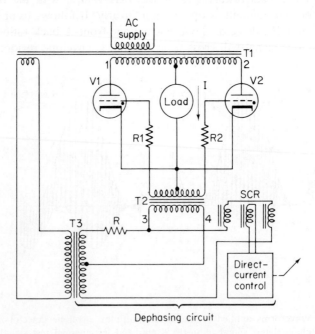

Fig. 5-28. A basic thyratron load-current control circuit.

reactor SCR, and the DC control unit. The purpose of the SCR is to provide a variable inductance, which operates in combination with R as an RL phase-shifting circuit. Thereby the phase of the voltage between terminals 3 and 4 of T2 can be shifted over a wide range. Note that the SCR operates as a variable inductor, because its core becomes more or less magnetically saturated as the value of the DC control current is varied.

Figure 5-29 shows normal operating waveforms for the circuit of Fig. 5-28. It is helpful to employ a dual-trace oscilloscope in this application, and to display the grid-cathode and anode-cathode waveforms simultaneously. As the DC control is varied through its range, the alternating grid voltages normally shift through the variable phase-shift angles, as indicated. Tubes V1 and V2 must be tested separately, of course. If normal waveforms and phase variation are found for V1, for example, and abnormal waveforms and/or phase variation are found for V2, preliminary trouble localization is thereby accomplished. Thus, R2 might be defective, or there might be a defect in the part of the secondary winding for T1 that supplies V2.

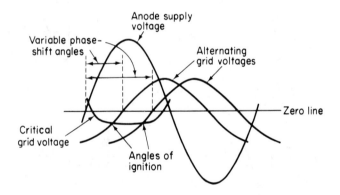

Fig. 5-29. Normal operating waveforms for the circuit in Fig. 5-28.

In the case that both of the alternating grid voltages are absent, the saturable-core reactor might be open-circuited. Or the primary winding of T2 might be open-circuited. As a practical note, apparent open-circuited windings are sometimes the result of poor terminal connections. If normal waveforms and phase variation are found for both V1 and V2, but the load current is weak or zero, the tubes may be defective. But if tube replacement does not correct the trouble, check the load (usually a DC motor) for normal operating condition. The practical troubleshooter must be on the alert for occasional production errors. For example, you have probably never encountered a motor nameplate stamped with an incorrect

135

current rating. However, this sort of "goof" can and does happen occasionally.

Note that an oscilloscope can be used to check the condition of a thyratron tube, without the necessity for procuring a new tube and making a substitution test. Figure 5-30 shows how an oscilloscope is connected between the anode and cathode of a thyratron, to check its operating waveform. If the cathode is not at ground potential, it is necessary to use an oscilloscope with double-ended vertical input. A normal operating waveform is depicted in (b). The arc drop will vary from 8 to 16 volts, depending on the tube type. An excessive arc drop, as exemplified in (c), shows that the tube is failing, is passing subnormal current, and is operating at poor efficiency. Another type of tube-operating fault is illustrated in (d). Here, there is an array of excessive arc drops that may be caused by the tube operating at a subnormal temperature, or by some comparable defect such as incorrect gas pressure.

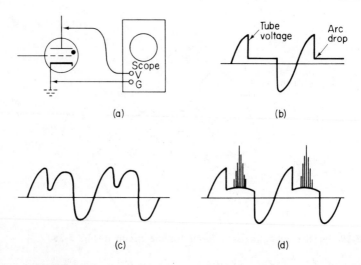

Fig. 5-30. Oscilloscope check of thyratron action. (a) Test setup (for thyratron with grounded cathode). (b) Normal operating waveform. (c) Failing tube with excessive arc drop. (d) Tube operating at too low a temperature, or equivalent fault.

Digital computers of various kinds are utilized in the more sophisticated and elaborate types of automated equipment. Circuit actions are basically switching operations performed by transistors or integrated circuits. These switching actions are activated by pulse sequences, as exemplified in Fig. 5-31. The essential points to be observed during signal-

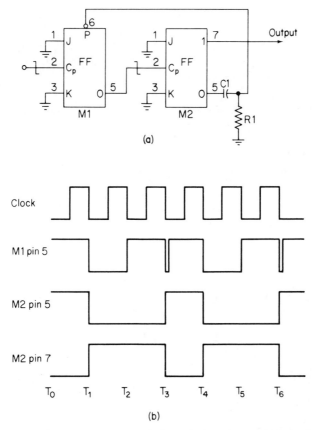

(a)

(b)

Fig. 5-31. Ideal pulse waveforms and timing relations for a divide-by-3 circuit. (a) Block diagram of flip-flop circuitry. (b) Normal waveforms at various test points. (*Courtesy of* Tektronix)

tracing procedures are pulse amplitudes, pulse widths, and relative timing of the pulses in the various circuit sections. In turn, a triggered-sweep oscilloscope is essential. Dual-trace displays help to speed up timing checks. Malfunction of digital-computer circuitry is often caused by component or device defects that distort the pulse waveforms excessively. Basic terminology employed in discussing pulse distortions is noted in Fig. 5-32.

Pulse distortions due to transistor characteristics are depicted in Fig. 5-33. The pulse-delay time t_d is the time delay that occurs after the application of an input pulse and a change in output equal to 10 per cent of the maximum change. This delay time is essentially the result of the junction capacitance between the emitter and the base, which requires time to charge, and also to the transit time of the initial emitter current in its

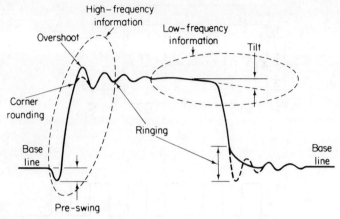

Fig. 5-32. Basic pulse-distortion terminology.

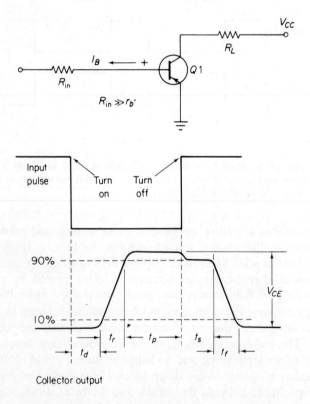

Fig. 5-33. Transistor switching characteristics.

138

diffusion through the base to the collector. The pulse-rise time t_r is the time required for the pulse to rise from 10 per cent to 90 per cent of its maximum amplitude. Pulse rise time is chiefly dependent upon the frequency response of the transistor, the amplitude of the applied base pulse, and the circuit capacitance. The duration of the maximum-amplitude portion of the pulse is denoted t_p. The storage time t_s is the result of diffusion of large numbers of minority carriers into the base region. This storage time is the interval during which the output amplitude remains near its maximum value after the input pulse is no longer present.

Accumulation of minority carriers into the base region occurs while the transistor is in its saturated state, and the base is forward-biased with

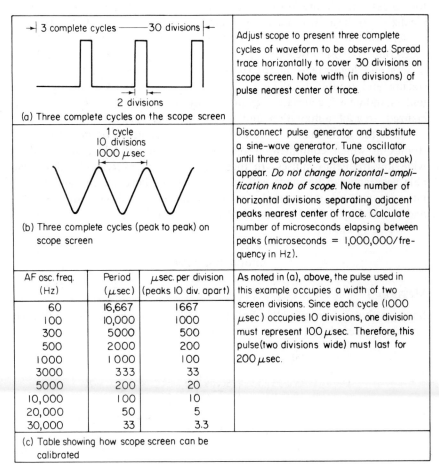

AF osc. freq. (Hz)	Period (μsec)	μsec. per division (peaks 10 div. apart)
60	16,667	1667
100	10,000	1000
300	5000	500
500	2000	200
1000	1000	100
3000	333	33
5000	200	20
10,000	100	10
20,000	50	5
30,000	33	3.3

(a) Three complete cycles on the scope screen

Adjust scope to present three complete cycles of waveform to be observed. Spread trace horizontally to cover 30 divisions on scope screen. Note width (in divisions) of pulse nearest center of trace.

(b) Three complete cycles (peak to peak) on scope screen

Disconnect pulse generator and substitute a sine-wave generator. Tune oscillator until three complete cycles (peak to peak) appear. *Do not change horizontal-amplification knob of scope.* Note number of horizontal divisions separating adjacent peaks nearest center of trace. Calculate number of microseconds elapsing between peaks (microseconds = 1,000,000/frequency in Hz).

(c) Table showing how scope screen can be calibrated

As noted in (a), above, the pulse used in this example occupies a width of two screen divisions. Since each cycle (1000 μsec) occupies 10 divisions, one division must represent 100 μsec. Therefore, this pulse (two divisions wide) must last for 200 μsec.

Fig. 5-34. Time measurement by means of a linear sweep—Method 1.

139

respect to both the emitter and collector. Storage delay occurs because, although the base signal would normally turn the transistor off, the minority carriers must be pulled from the base region into the collector circuit before the transistor can change its state. This process is followed by the fall-time interval t_f, which is analogous to the rise-time interval. It is due chiefly to the time that is required for the junction capacitance between the collector and base to discharge. In summary, the chief waveform distortions due to device characteristics are a slow-down of the rise and fall times, and a delay of the output pulse with respect to the input pulse.

There are permissible tolerances on pulse distortion. When device defects or component faults cause pulse distortion to exceed normal tolerances, circuit malfunctions result. This type of trouble is generally localized to best advantage by means of signal-tracing tests. Because modern digital computers operate at extremely high speed, conventional industrial-type oscilloscopes are inadequate. Very wide-band vertical amplifiers and high-precision triggered sweeps are essential. Effective signal tracing and trouble localization are also dependent upon the operator's understanding of computer operation. In turn, this is a somewhat specialized area of automation and industrial electronics.

On the other hand, there are various other areas of industrial electronics in which equipment is operated at comparatively slow speed by wide low-frequency pulses. In these situations, measurements of reasonable accuracy can be made with an ordinary oscilloscope and audio oscillator. A worked-out example of the procedure is shown in (a), (b), and (c) of Fig. 5-34. Procedural details are as follows:

Information Sheet

METHOD 1

Suppose that it is required to find the width, in microseconds, of the rectangular pulses shown at (a) in Fig. 5-34.

1. Set up the equipment as depicted in Fig. 5-35.

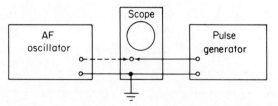

Fig. 5-35. Block diagram of equipment used in measuring pulse width.

2. Display on the CRT screen the pulse waveform to be measured. Proceed as follows:

(a) Connect the source of the pulses to the vertical-input terminals of the oscilloscope.

(b) Adjust the operating controls to display three cycles on the screen.

Note: Do not change the horizontal gain-control setting after making this adjustment. To do so would change the calibration of the scope, and make the readings meaningless.

(c) Note and record the number of horizontal screen divisions occupied by the pulse to be measured.

3. Calibrate the oscilloscope as follows:

(a) Disconnect the pulse-generator leads from the vertical-input terminals of the oscilloscope, and apply an audio-oscillator output signal.

(b) Tune the audio oscillator to display three sine-wave cycles on the CRT screen. The audio oscillator is now operating at the same frequency as the pulse generator. (Remember—do not change the horizontal gain-control setting of the oscilloscope.)

(c) Determine the number of microseconds represented by each screen division. For example, if the frequency dial of the audio oscillator reads 1 kHz, then each cycle has a period of 0.001 second, or 1000 microseconds. Suppose that the distance between two adjacent sine-wave peaks is found to be 10 divisions. In turn, this indicates that the 10 divisions represent 1000 microseconds, and that one division represents 100 microseconds.

4. Measure the width of the pulses as follows:

(a) Refer to your notes, and determine the width (in divisions) of the displayed pulses, as measured in step 2.

(b) Multiply this number of divisions by the number of microseconds represented by one pulse division. For example, suppose that the pulse was found to occupy two divisions in width. Knowing that each division represents 100 microseconds, it follows that the pulse has a width of 200 microseconds (2 divisions × 100 microseconds).

In a similar manner, the time elapsing between any two points on a pulse can be measured. It is first necessary to note how many horizontal divisions separate the two points whose time difference is to be measured. Then, multiply this number of divisions by the number of microseconds represented by each division.

The measurement can be no more accurate than the calibration accuracy of the audio-oscillator frequency dial.

METHOD 2

Another method of time measurement that can be used on low-frequency pulse waveforms is shown in Fig. 5-36. The procedure is as follows:

1. Connect the equipment as shown in Fig. 5-35.

2. Display one or more pulses on the CRT screen, and expand the pattern to any convenient width. Observe and record the number

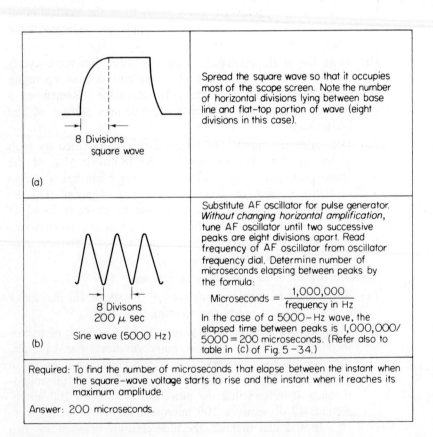

8 Divisions square wave (a)	Spread the square wave so that it occupies most of the scope screen. Note the number of horizontal divisions lying between base line and flat-top portion of wave (eight divisions in this case).
8 Divisons 200 μ sec Sine wave (5000 Hz) (b)	Substitute AF oscillator for pulse generator. *Without changing horizontal amplification*, tune AF oscillator until two successive peaks are eight divisions apart. Read frequency of AF oscillator from oscillator frequency dial. Determine number of microseconds elapsing between peaks by the formula: $$\text{Microseconds} = \frac{1,000,000}{\text{frequency in Hz}}$$ In the case of a 5000-Hz wave, the elapsed time between peaks is 1,000,000/5000 = 200 microseconds. (Refer also to table in (c) of Fig. 5-34.)

Required: To find the number of microseconds that elapse between the instant when the square-wave voltage starts to rise and the instant when it reaches its maximum amplitude.

Answer: 200 microseconds.

Fig. 5-36. Time measurement by means of a linear sweep—Method 2.

of horizontal divisions occupied by the pulse. Do not change the setting of the horizontal gain control thereafter.

3. Disconnect the pulse generator from the oscilloscope and connect an audio oscillator in its place.

4. Adjust the frequency dial of the audio oscillator until adjacent sine-wave peaks are separated by exactly the same number of divisions measured in step 2.

5. Calculate how many microseconds elapse between adjacent peaks, according to the formula

$$\text{Microseconds} = \frac{1,000,000}{\text{Frequency in Hertz}}$$

Waveshaping Experiment

SUBJECT

Signal tracing and waveform analysis in a waveshaping chain.

OBJECTIVE

To show how a sine wave can be processed into a series of narrow pulses of either positive or negative polarity. To accustom the oscilloscope operator to system operation when several waveshaping circuits are connected together in cascade.

MATERIAL REQUIRED

1. Triode limiter chassis (Fig. 5-37).

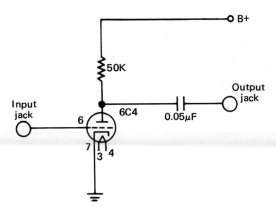

Fig. 5-37. Schematic of a triode limiter.

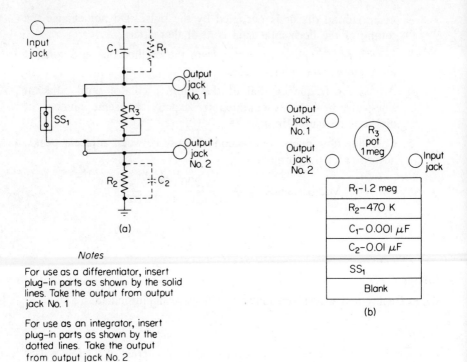

Notes

For use as a differentiator, insert plug-in parts as shown by the solid lines. Take the output from output jack No. 1

For use as an integrator, insert plug-in parts as shown by the dotted lines. Take the output from output jack No. 2

Fig. 5-38. Differentiator-integrator chassis. (a) Schematic. (b) Bottom view.

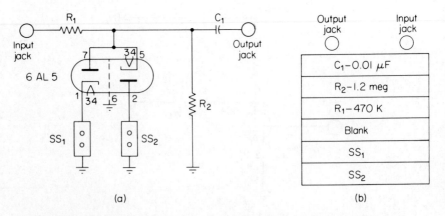

Fig. 5-39. Parallel diode limiter chassis. (a) Schematic. (b) Bottom view.

2. Differentiator-integrator chassis (Fig. 5-38).

3. Parallel diode limiter chassis (Fig. 5-39).

4. Bias supply chassis (Fig. 5-40).

5. Service-type oscilloscope.

6. Bench power supply.

7. Audio oscillator with leads.

8. VTVM or TVM with leads.

INSTRUCTIONS

In industrial control equipment, it may be necessary to convert a sine wave into a series of narrow, accurately timed pulses. This experiment shows a basic method of producing pulses that have accurate spacing. It

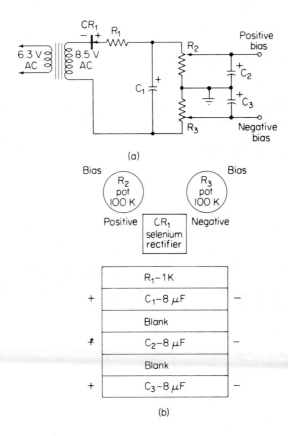

(a)

(b)

Fig. 5-40. Bias supply chassis. (a) Schematic. (b) Bottom view.

145

also serves to introduce the oscilloscope operator to a basic example of system operation.

Procedure

The Overdriven Amplifier

1. Connect the equipment as shown in Fig. 5-41.

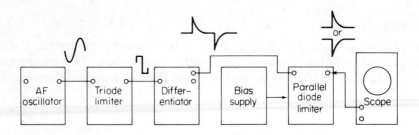

Fig. 5-41. Block diagram of the wave-shaping chain.

2. Observe the waveform being applied by the audio oscillator. Sketch this waveform at *A* in Fig. 5-42.

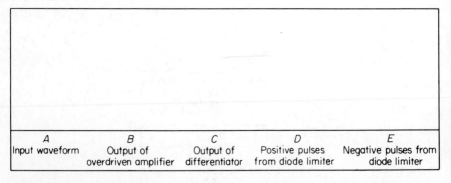

Fig. 5-42. Typical waveforms encountered in wave-shaping circuits.

3. Connect the oscilloscope to the output of the triode limiter, and adjust the limiter for an approximate square-wave output waveform. Sketch this waveform at *B* in Fig. 5-42.

The Differentiator

1. Calculate a suitable set of values for C and R in the differentiator

chassis. Remember that for good differentiation, the following conditions should be observed:

(a) The RC time constant must be very short, compared with the period of the waveform to be differentiated.

(b) R must be large, as compared with the plate load resistance of the preceding stage, to avoid undue loading of the driving circuit.

2. Connect the RC value that you have calculated into the differentiator circuit.

3. Record the displayed waveform at C in Fig. 5-42.

The Parallel Diode Limiter

1. Set up the parallel diode limiter chassis to produce only positive pulses at the output jack.

2. Observe the waveform displayed at the output jack, and sketch the waveshape at D in Fig. 5-42.

3. Set up the parallel diode limiter chassis to produce only negative pulses.

4. Observe the waveform displayed at the output jack, and sketch its waveshape at E in Fig. 5-42.

5. Measure the time in microseconds that elapses between two successive negative pulses at the output of the parallel diode limiter chassis. Record this value.

6. Measure the width of the pulses (at the base line) in microseconds. Record this value.

7. Adjust the negative pulses to an amplitude of 10 volts, peak. Ask your instructor to check the peak value of the displayed waveform.

Conclusions

1. Draw a block diagram of a wave-shaping chain capable of converting a sine wave into the waveform shown in the square below. Show the input and output voltage waveforms for each stage.

2. In the same way, draw a block diagram of a chain that will develop the waveform shown in the square below.

3. Draw a block diagram of a chain that will convert a sine wave into the waveform shown in the square below.

6

TV TROUBLESHOOTING
WITH THE
OSCILLOSCOPE

6.1 GENERAL CONSIDERATIONS

TV troubleshooting with the oscilloscope involves the circuit sections depicted in Fig. 6-1. Tests of the tuner section include frequency-response, source-voltage ripple, and AGC spurious-voltage tests. Figure 6-2 shows the plan of a sweep-frequency (visual-alignment) test. A sweep-frequency generator applies a frequency-modulated signal to the tuned circuits under test. This FM signal varies in frequency over the pass band of the tuned circuits at a 60-Hz repetition rate. In turn, a frequency response curve is displayed on the CRT screen. This curve is basically a graph of output voltage vs. frequency. Note that a detector is employed to demodulate the output signal before it is applied to the oscilloscope. In many cases, this detector will be a part of the tuned-circuit section under test.

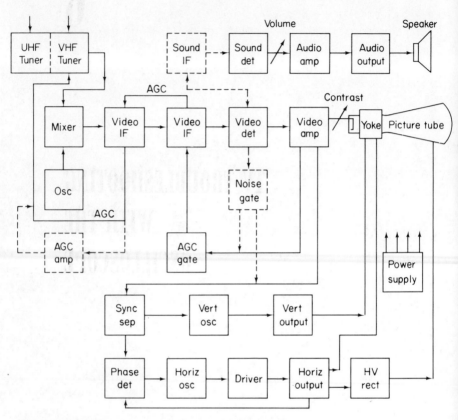

Fig. 6-1. Block diagram of a typical television receiver.

6.2 VHF AND UHF TUNER TESTS

A test setup for checking the frequency response of a VHF tuner is shown in Fig. 6-3, along with a typical frequency-response curve. Note that the VHF sweep signal is applied to the antenna-input terminals of the tuner. The oscilloscope is connected through a 50-k isolating resistor to the "looker point" on the tuner. Since the "looker point" is a tap in the mixer circuit (heterodyne section), demodulator (detector) action occurs in the tuner. The isolating resistor is employed to provide low-pass filter action and thereby sharpen beat-marker indications on the response curve. Most sweep generators contain built-in marker generators and a properly phased 60-Hz sine-wave horizontal-deflection voltage for the oscilloscope. However, the scope can be operated on 60-Hz sawtooth deflection, if desired. Note that the AGC line is clamped with a fixed value of bias voltage, as

150

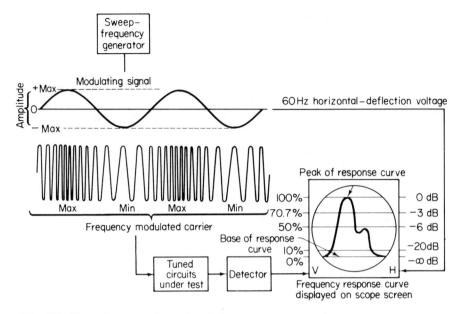

Fig. 6-2. Plan of a sweep-frequency test.

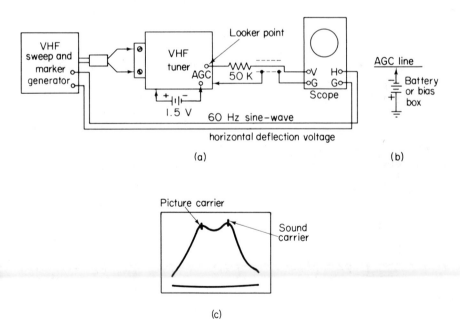

Fig. 6-3. Test setup for a VHF-tuner frequency-response test. (a) Test setup. (b) Fixed-bias clamping of AGC line. (c) Typical frequency-response curve.

151

specified in the receiver service data. Fixed bias stabilizes the development of the frequency response curve.

Channel no.	Picture carrier (P)	Sound carrier (S)	Freq. limits
2	P 55.25	S 59.75	54 – 60
3	P 61.25	S 65.75	60 – 66
4	P 67.25	S 71.75	66 – 72
5	P 77.25	S 81.75	76 – 82
6	P 83.25	S 87.75	82 – 88
7	P 175.25	S 179.75	174 – 180
8	P 181.25	S 185.75	180 – 186
9	P 187.25	S 191.75	186 – 192
10	P 193.25	S 197.75	192 – 198
11	P 199.25	S 203.75	198 – 204
12	P 205.25	S 209.75	204 – 210
13	P 211.25	S 215.75	210 – 216
14	P 471.25	S 475.75	470 – 476
15	P 477.25	S 481.75	476 – 482
16	P 483.25	S 487.75	482 – 488
17	P 489.25	S 493.75	488 – 494
18	P 495.25	S 499.75	494 – 500
19	P 501.25	S 505.75	500 – 506
20	P 507.25	S 511.75	506 – 512
21	P 513.25	S 517.75	512 – 518
22	P 519.25	S 523.75	518 – 524
23	P 525.25	S 529.75	524 – 530
24	P 531.25	S 535.75	530 – 536
25	P 537.25	S 541.75	536 – 542
26	P 543.25	S 547.75	542 – 548
27	P 549.25	S 553.75	548 – 554
28	P 555.25	S 559.75	554 – 560
29	P 561.25	S 565.75	560 – 566
30	P 567.25	S 571.75	566 – 572
31	P 573.25	S 577.75	572 – 578
32	P 579.25	S 583.75	578 – 584
33	P 585.25	S 589.75	584 – 590
34	P 591.25	S 595.75	590 – 596
35	P 597.25	S 601.75	596 – 602
36	P 603.25	S 607.75	602 – 608
37	P 609.25	S 613.75	608 – 614
38	P 615.25	S 619.75	614 – 620
39	P 621.25	S 625.75	620 – 626
40	P 627.25	S 631.75	626 – 632
41	P 633.25	S 637.75	632 – 638
42	P 639.25	S 643.75	638 – 644
43	P 645.25	S 649.75	644 – 650
44	P 651.25	S 655.75	650 – 656
45	P 657.25	S 661.75	656 – 662
46	P 663.25	S 667.75	662 – 668
47	P 669.25	S 673.75	668 – 674
48	P 675.25	S 679.75	674 – 680
49	P 681.25	S 685.75	680 – 686
50	P 687.25	S 691.75	686 – 692
51	P 693.25	S 697.75	692 – 698
52	P 699.25	S 703.75	698 – 704
53	P 705.25	S 709.75	704 – 710
54	P 711.25	S 715.75	710 – 716
55	P 717.25	S 721.75	716 – 722
56	P 723.25	S 727.75	722 – 728
57	P 729.25	S 733.75	728 – 734
58	P 735.25	S 739.75	734 – 740
59	P 741.25	S 745.75	740 – 746
60	P 747.25	S 751.75	746 – 752
61	P 753.25	S 757.75	752 – 758
62	P 759.25	S 763.75	758 – 764
63	P 765.25	S 769.75	764 – 770
64	P 771.25	S 775.75	770 – 776
65	P 777.25	S 781.75	776 – 782
66	P 783.25	S 787.75	782 – 788
67	P 789.25	S 793.75	788 – 794
68	P 795.25	S 799.75	794 – 800
69	P 801.25	S 805.75	800 – 806
70	P 807.25	S 811.75	806 – 812
71	P 813.25	S 817.75	812 – 818
72	P 819.25	S 823.75	818 – 824
73	P 825.25	S 829.75	824 – 830
74	P 831.25	S 835.75	830 – 836
75	P 837.25	S 841.75	836 – 842
76	P 843.25	S 847.75	842 – 848
77	P 849.25	S 853.75	848 – 854
78	P 855.25	S 859.75	854 – 860
79	P 861.25	S 865.75	860 – 866
80	P 867.25	S 871.75	866 – 872
81	P 873.25	S 877.75	872 – 878
82	P 879.25	S 883.75	878 – 884
83	P 885.25	S 889.75	884 – 900

P = Picture carrier frequency   S = Sound carrier frequency   All frequencies in MHz

Fig. 6-4. Picture and sound carrier frequencies for the VHF and UHF TV channels. (*Courtesy of* Howard W. Sams & Co., Inc.)

There are 12 VHF television channels, as tabulated in Fig. 6-4. Each pair of picture and sound carriers is separated by 4.5 MHz. As depicted in Fig. 6-3, the picture and sound carriers normally fall on the peaks of the VHF response curve. However, due to manufacturing tolerances, it is sometimes impossible to align the tuner circuits for a double-humped response. This aspect of curve tolerance is shown in Fig. 6-5. If the curve exceeds tolerance limits of +15 and −30 per cent, it is concluded that there is circuit trouble present, and that the tuner requires troubleshooting before it can be properly aligned. Another aspect of alignment tolerance is the necessity for compromise adjustments in some tuner designs. That is, an alignment adjustment on one channel may affect the alignment adjustment on another channel. In such a case, compromise adjustments are made so that the curves on all channels are within tolerance insofar as possible.

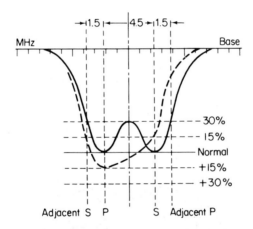

Fig. 6-5. Tolerance limits for a VHF frequency-response curve.

A UHF tuner is sweep-aligned through the VHF tuner, as shown in Fig. 6-6. The resulting overall UHF/VHF response curve normally appears essentially the same as the VHF response curve (Fig. 6-3). If the overall curve is out of tolerance limits of +15 and −30 percent (Fig. 6-5), it is concluded that there is circuit trouble present, and that the UHF tuner requires troubleshooting before it can be properly aligned. Note that repair of VHF and UHF tuners is a somewhat specialized activity, and that many technicians send defective tuners to central repair depots.

In addition to the foregoing tests, an oscilloscope is also used to check the source-voltage ripple amplitude. This test is made at the source-

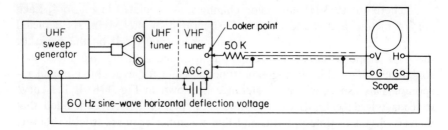

Fig. 6-6. Test setup for a UHF-tuner frequency-response check.

voltage terminal on the tuner. The ripple amplitude is normally less than 1 per cent of the source-voltage value. A typical ripple waveform is illustrated in Fig. 6-7. Abnormally high ripple is caused by defective filter capacitors in the power supply, or by defective decoupling capacitors in the vertical-sweep or vertical-oscillator sections. Excessive ripple voltage causes hum bars in the picture, picture pulling, and sometimes complete loss of horizontal synchronization.

The AGC line is normally free from any ripple voltage or spurious AC voltages. However, if the AGC bypass capacitors become open, an oscilloscope will show the presence of 15,750-Hz pulses and other AC waveforms on the AGC line. These spurious AC voltages cause contrast variations in various portions of the picture. High-level spurious voltages can also cause picture pulling and erratic synchronizing action.

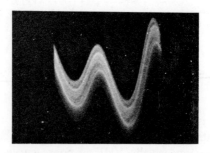

Fig. 6-7. A typical ripple waveform.

6.3 IF AMPLIFIER TESTS

Oscilloscope tests in the IF-amplifier section comprise signal-tracing and frequency-response checks. Signal tracing is accomplished by means of a demodulator probe, such as depicted in Fig. 6-8(a). The probe must be

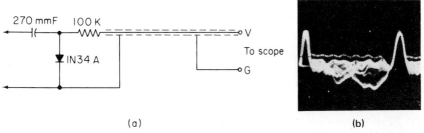

(a)   (b)

Fig. 6-8. A demodulator probe is used to signal trace in the IF section. (a) Typical demodulator-probe configuration. (b) Vertical sync pulse is the most prominent demodulated waveform component.

used with a sensitive oscilloscope in order to obtain useful indications in the first IF stage. A vertical amplifier with a sensitivity of 10 millivolts rms per inch is adequate. The receiver under test must be tuned to a TV station, or driven by a test-pattern generator. Note that the oscilloscope should be operated on 60-Hz horizontal deflection rate. This requirement is imposed by the comparatively limited demodulation ability of the probe. That is, the vertical sync pulse is the most prominent component in the demodulator-probe output waveform, as seen in Fig. 6-8(b).

If the oscilloscope is operated at 15,750-Hz horizontal deflection rate, it is impossible to see the vertical sync-pulse component of the waveform. Instead, only the horizontal sync-pulse component is visible. Since the horizontal sync pulse is greatly attenuated by the demodulator probe, this is not a practical mode of oscilloscope operation. Note that a demodulator probe tends to load and detune the various IF circuits, with the result that waveform amplitude measurements are generally meaningless. However, a signal-tracing test is very useful to determine whether a signal is present or absent at a particular point in the IF amplifier. In case there is no signal present, it is indicated that there is either a component defect in the circuit branch under test, or that there is a defect in the AGC system that is biasing off the stage.

An IF frequency-response test setup is depicted in Fig. 6-9. The output from an IF sweep-and-marker generator is applied to the base of the mixer transistor through a 0.001-μf capacitor. This series capacitor prevents base-bias drain-off through the generator cable. The IF AGC line is clamped to a fixed voltage specified in the receiver service data. Clamping prevents the AGC system from reacting to the sweep signal and possibly distorting the response curve. Note that the oscilloscope is connected across the video-detector load resistor. A 100-k series-isolating resistor is employed to provide low-pass filter action and sharpen the marker indication. Horizontal deflection of the CRT beam

155

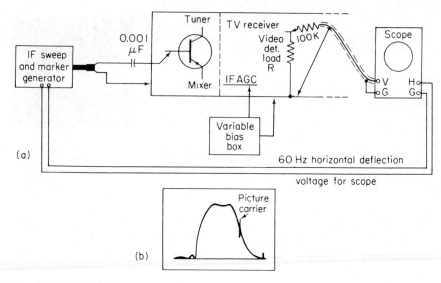

(a)

(b)

Fig. 6-9. Typical IF frequency-response test setup. (a) Equipment connections. (b) IF response-curve display.

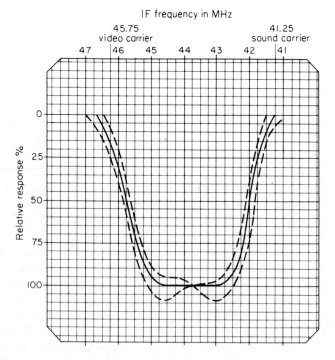

Fig. 6-10. IF response-curve tolerance limits.

156

is usually controlled by a properly phased 60-Hz sine-wave voltage provided by the sweep generator.

Note that the picture (video) carrier falls halfway down the side of
an IF response curve, at 45.75 MHz. The sound carrier is 4.5 MHz
below the picture carrier, at 41.25 MHz. Figure 6-10 shows the relation of
the picture and sound carriers, and also the response-curve tolerance
limits for a typical receiver. Note also that the frequency progression in
a displayed IF response curve may increase from left to right, or from
right to left, depending upon the instruments that are utilized. That is,
the aspect of the IF curve depends upon the phase of the horizontal-
deflection voltage for the oscilloscope. Again, a response curve may be
displayed "right side up" or "upside down." (See Fig. 6-11.) This
aspect depends upon the polarity of the video-detector diode in the receiver under test. The bandwidth of a response curve is defined as the
number of MHz between its 50 per cent amplitude points. In the example
of Fig. 6-10, the curve bandwidth is 3.75 MHz.

Fig. 6-11. Response curve may be displayed "right side up" or "upside down."

6.4 VIDEO AMPLIFIER TESTS

Oscilloscope tests in the video-amplifier section include signal-tracing,
gain-measurement, square-wave, and frequency-response checks. Signal
tracing is accomplished with a service-type oscilloscope and low-capacitance probe. The probe minimizes circuit loading in the amplifier. If the
receiver is energized by a TV station signal, and the oscilloscope is deflected at a 7875-Hz rate, composite video waveforms with positive-going
or negative-going horizontal sync pulses are normally displayed, as illustrated in Fig. 6-12. The polarity of the display depends upon the polarity
of the video detector, and also upon the test point in the video amplifier
at which the probe is applied. For example, if a positive-going pulse is
displayed at the base of a transistor, a negative-going pulse will be displayed at the collector in the CE configuration.

Gain measurements are made by comparison of the waveform amplitudes at the input and at the output of a stage. As an illustration, if the
output waveform has an amplitude ten times as great as the input wave-

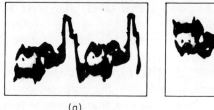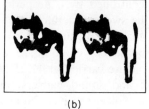

(a)   (b)

Fig. 6-12. Display of the composite video waveform. (a) Positive-going sync pulses.
(b) Negative-going sync pulses.

form, the voltage gain of the stage is ten times. Gain measurements are
made to best advantage by energizing the receiver with a generator pattern
signal, instead of utilizing a TV station signal. A generator signal is
steady, whereas a station signal fluctuates in amplitude. If the waveform
at the video-detector output has a subnormal peak-to-peak voltage with
respect to the specification in the receiver service data, the trouble may
be found in either the input circuit of the first video-amplifier stage, or
in the sections prior to the video amplifier. For example, the video-
detector diode might have a poor front-to-back ratio, or there might be
an IF-amplifier defect, or a tuner defect.

Square-wave tests in video amplifiers are usually made at a repetition
rate of 100 kHz, with the test setup depicted in Fig. 6-13. It is advisable
to disconnect the picture-detector diode during the test, to avoid input-
loading variations. An oscilloscope and low-capacitance probe are connected at the output of the video amplifier. Note that the oscilloscope must
have faster rise than the video amplifier in order to obtain a valid test.
A rise time of 0.08 μs is adequate. Similarly, the square-wave generator
must apply a waveform with a rise time of 0.08 μs. Otherwise, the reproduced square wave will be slowed down more or less by the instrument
characteristics. A typical video amplifier has a rise time of 0.1 μs. An
overshoot and/or undershoot of 10 per cent is permissible. (See Fig.
6-14.)

When the square-wave response of a video amplifier is unsatisfactory,
the trouble may be caused by an open decoupling capacitor, a load resistor that has increased substantially in value, or a defective peaking coil.
With reference to Fig. 6-15, the square-wave rise time will be greatly
slowed if C_F becomes open-circuited. The same effect is produced in
case R_L increases substantially in resistance. Distorted square-wave response can also be caused by short circuits or partial short circuits in
peaking coils L_1 and L_2. Note that peaking-coil inductance values are
comparatively critical, and that an incorrect replacement coil can lead to
baffling trouble symptoms.

158

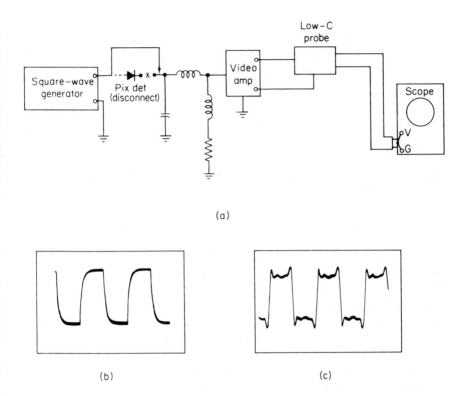

(a)

Fig. 6-13. Square-wave check of video-amplifier response. (a) Test setup. (b) Output from medium-performance amplifier. (c) Output from high-performance amplifier.

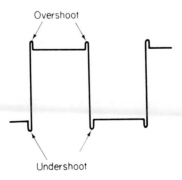

Fig. 6-14. Example of 10 per cent overshoot and undershoot.

159

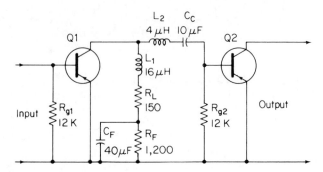

Fig. 6-15. A video-amplifier configuration with frequency response from 30 Hz to 4 MHz.

There is a useful relation between the rise time of a reproduced square wave and the bandwidth of an amplifier. This relation states that the rise time is equal to ⅓ of the period for the high cutoff frequency of the amplifier. As an illustration, an amplifier with frequency response to 4 MHz will have a rise time of 0.083 μs. That is, the period corresponding to 4 MHz is 0.25 μs, and ⅓ of this period is equal to 0.083 μs. Note that the bandwidth of a video amplifier is defined as the number of MHz between the 50 per cent amplitude points, as in the case of an IF frequency-response curve.

There is also a useful relation between the tilt of a reproduced square wave and the low-frequency cutoff point of the amplifier. Figure 6-16 shows the meaning of tilt; it is the difference between amplitudes E_2 and E_1. A comparatively low square-wave repetition rate is utilized when one is measuring tilt—the rep rate should be reduced until a tilt of 10 or 15 per cent is observed. The relation between tilt and low-frequency cutoff is written:

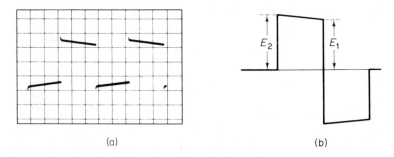

(a)   (b)

Fig. 6-16. Tilt in a reproduced square wave. (a) Waveform display. (b) Tilt measurement.

$$f_c = \frac{2f(E_2 - E_1)}{3(E_2 + E_1)}$$

where f_c is the low cutoff frequency.
  f is the square-wave frequency.
  E_2 and E_1 are the amplitudes shown in Fig. 6-16.

As an illustration, let us suppose that we measure 15 per cent tilt in a reproduced 60-Hz square wave. That is, $E_2 - E_1 = 0.15$. In turn, the low cutoff frequency is approximately 3 Hz.

A frequency-response check of a video amplifier is made as shown in Fig. 6-17. The output from a video-frequency sweep generator is applied to the input of the video amplifier. A demodulator probe is connected to the output of the video amplifier. In turn, the video-frequency response curve is displayed. A bandwidth from 3.5 to 4 MHz is typical. Distorted response and/or subnormal bandwidth can result from the same

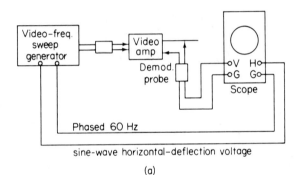

(a)

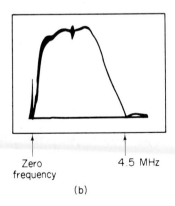

(b)

Fig. 6-17. Checking video-amplifier frequency response. (a) Test setup. (b) Typical response curve.

defects that cause unsatisfactory square-wave response. Frequency-response curves for video amplifiers are seldom specified in service data for black-and-white TV receivers. However, when one is analyzing a defective video amplifier, it is usually possible to make a comparison check with another receiver that is in normal operating condition.

6.5 SYNC CHANNEL TESTS

Oscilloscope tests in the sync channel consist of waveform analysis and peak-to-peak voltage measurements. The sync tips are clipped from the

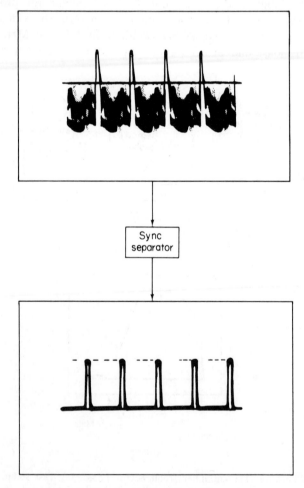

Fig. 6-18. Key waveforms in the sync-separator section.

composite video signal in the sync-separator sections, as shown in Fig. 6-18. A single transistor or diode may be employed as a clipper in economy-type receivers. Deluxe receivers utilize a transistor and a diode, or two transistors. The chief requirement for a sync separator is provision of a stripped-sync pulse output free from residual video signal. Figure 6-19 exemplifies contaminated and clean stripped-sync waveforms. Correct clipping action requires precise biasing of the clipping device. Incorrect bias voltage is commonly caused by leaky fixed capacitors. Collector leakage in a clipper transistor, or a poor front-to-back ratio in a clipper diode, can also cause faulty sync separation.

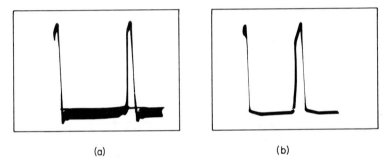

(a)   (b)

Fig. 6-19. Sync-separator output waveforms. (a) Spurious video signal present. (b) All video signal eliminated.

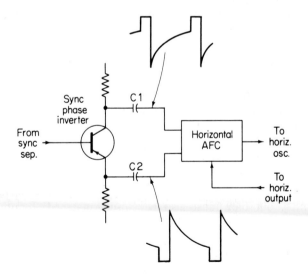

Fig. 6-20. Arrangement of a sync phase inverter.

Many receivers employ a sync phase inverter, as depicted in Fig. 6-20. This section changes the pulse output from the sync separator into a push-pull waveform for operation of the horizontal-AFC section. These inverter output waveforms should have equal amplitudes as specified in the receiver service data. Incorrect waveforms are often caused by leaky or open capacitors, such as C1 and C2. Note that inverter output waveforms often have a sawtooth component in addition to a pulse component. This sawtooth component results from the injection of a comparison waveform from the horizontal-output section into the horizontal-AFC section.

6.6 HORIZONTAL AFC TESTS

A typical horizontal-AFC configuration is shown in Fig. 6-21, with its operating waveforms. Oscilloscope tests in the AFC section include waveform analysis and peak-to-peak voltage measurements. Note that positive-going sync pulses are applied to the junction of R4 and R5. Two sawtooth comparison waveforms are employed. These sawtooth waveforms are obtained from the horizontal-output section through a phase splitter (phase inverter), so that a positive-going sawtooth is applied to D1 and a negative-going sawtooth is applied to D2. Note that the sync pulses fall part way down the flyback interval of the sawtooth waveforms. In normal operation, the peak voltages of W1 and W2 are equal.

Next, let us suppose that the horizontal oscillator tends to run too slowly. In turn, the pulse and sawtooth waveforms are no longer exactly in phase with each other. As depicted in Fig. 6-21, the pulse rides higher on the sawtooth in W1, and rides lower on the sawtooth in W2. Accordingly, the peak voltage of W1 now exceeds the peak voltage of W2. In turn, diode D1 conducts more than diode D2, and a positive DC control voltage is developed. This control voltage speeds up the horizontal oscillator and brings the sawtooth wave back in phase with the pulse. Conversely, if the horizontal oscillator tends to run too fast, D2 conducts more than D1, and a negative control voltage is developed.

Incorrect AFC waveshapes and/or peak-to-peak voltages are usually caused by defective capacitors. Faulty diodes are also common troublemakers. Note that the diodes must not only have good front-to-back resistance ratios, but they must also be closely matched. In other words, the forward-resistance values of the diodes must be practically equal. Otherwise, one diode will conduct more than the other when the waveforms are normal, and the AFC circuit will be unbalanced. In turn, the control range will be impaired, or sync lock may be completely lost. Because of the balanced operating requirement, fixed-resistor values in

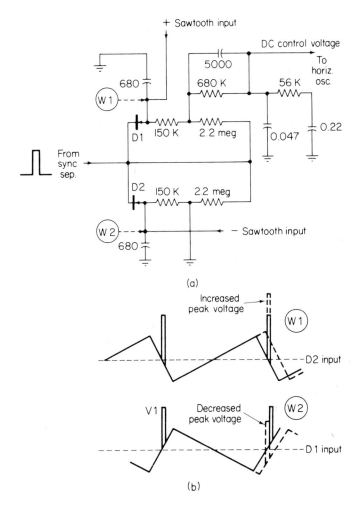

Fig. 6-21. Operation of a horizontal-AFC circuit. (a) Circuit arrangement. (b) Operating waveforms.

the diode circuit branches are also more critical than in various other circuits.

6.7 HORIZONTAL OSCILLATOR TESTS

Oscilloscope tests in the horizontal-oscillator section consist of waveform analysis and peak-to-peak voltage measurements, as specified in the receiver service data. Most horizontal oscillators employ a ringing coil, as

165

shown in Fig. 6-22, for stabilization of the operating frequency (reduction of jitter). A blocking oscillator configuration is utilized, with positive feedback via L1. Blocking action generates a spike output waveform, which shock-excites the ringing coil L2. In turn, L2 generates a sine-wave voltage, and the complete output waveform consists of a spike and a sine-wave component. For maximum stabilization of the oscillating action, the sine-wave component must be phased with respect to the spike component as illustrated in Fig. 6-22(d). This phase adjustment is controlled by the setting of the core slug in L2. Incorrect slug adjustments result in waveforms as exemplified in (b) and (c).

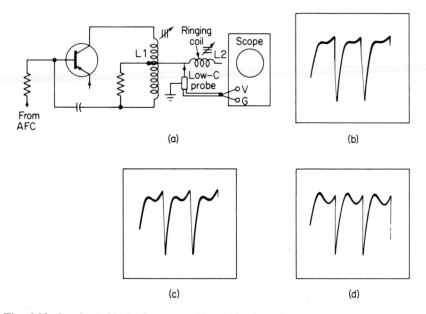

Fig. 6-22. Synchroguide horizontal-oscillator circuit and operation. (a) Test setup. (b–c) Incorrect waveforms. (d) Correct waveform.

6.8 VERTICAL OSCILLATOR TESTS

A block diagram for a vertical-oscillator section is depicted in Fig. 6-23. Oscilloscope tests include waveform analysis and peak-to-peak voltage measurements. A low-capacitance probe is used to minimize circuit loading. Note that the pulse output from the integrator cannot be properly displayed unless the vertical oscillator is temporarily disabled. That is, the oscillator waveform tends to back up into the integrator and distorts

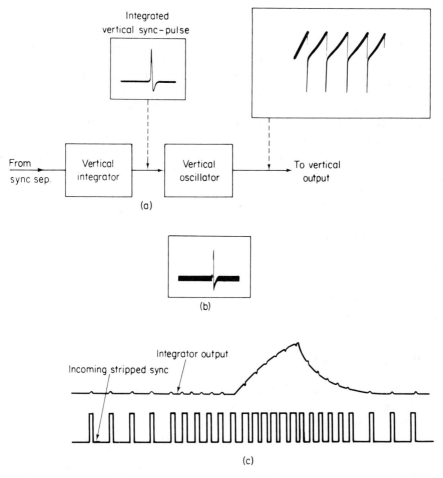

Fig. 6-23. Block diagram of vertical-oscillator section. (a) Functional sections and waveforms. (b) Faulty integrator output. (c) Expanded vertical sync pulse.

the pulse output from the integrator. The vertical oscillator can be disabled by shunting a large capacitor from the collector of the vertical-oscillator transistor to ground. If the integrated vertical-sync pulse has subnormal peak-to-peak voltage, the most likely cause is a leaky capacitor in the integrator circuit. Or, if baseline interference appears in the pulse display, as illustrated in Fig. 6-23(b), the most likely cause is an open capacitor in the integrator circuit.

Subnormal pulse amplitude impairs vertical sync lock, with the result that the picture is likely to break sync from time to time, and "roll"

167

on the screen. Of course, if there is a short-circuited capacitor in the integrator circuit, there will be no pulse output, and the picture will roll continuously. Interference along the base line of the integrated vertical-sync pulse consists of residual horizontal sync pulses. These residual pulses tend to cause loss of interlace (line pairing) with resulting impairment of picture detail. Pairing results from the fact that the incoming horizontal sync pulses have alternate half- and full-line spacing from the vertical sync pulse. In turn, triggering of the vertical oscillator occurs a half-line early on every other scan.

Note in Fig. 6-23 that the output waveform from the vertical oscillator is a peaked-sawtooth pattern. Its peak-to-peak voltage should be practically the same as specified in the receiver service data, and the proportions of the sawtooth and peaking-pulse components should also be correct. In this example, the peaking pulse has the same amplitude as the sawtooth component. Incorrect component proportions cause vertical scanning nonlinearity. The sawtooth ramp should also be approximately linear. Some degree of curvature in the ramp can be corrected by adjustment of the vertical-linearity control in the receiver. On the other hand, excessive ramp curvature cannot be corrected, and will cause vertical scanning nonlinearity.

6.9 OTHER RECEIVER WAVEFORMS

In addition to the waveforms that have been discussed in the foregoing topics, there are other receiver waveforms encountered in the sound IF section, audio section, horizontal driver section, horizontal-output section, vertical-output section, AGC section, and noise-gate section. However, the same principles are observed when these waveforms are checked, as noted in the foregoing topics. As an illustration, the sound-IF section is comparable to the video-IF section, except that it operates at a lower frequency and with less bandwidth. In turn, the same general principles apply in checking the alignment of the sound-IF section. The audio amplifier is checked in the same manner as explained for hi-fi amplifiers in the previous chapter. Note, however, that most TV audio amplifiers do not measure up to hi-fi requirements.

There are occasional points in TV receiver circuitry that are not checked with an oscilloscope. For example, the high-voltage section is avoided. Although it is possible to measure the ripple in the high-voltage supply, a special high-voltage probe is required. High-voltage probes are utilized in laboratories, but very seldom in service shops. If an attempt is made to check the high-voltage ripple with a low-capacitance probe, the probe and the input section of the oscilloscope will be damaged. In tube-type receivers, the voltage across the horizontal-deflection coils and in the damper circuit exceeds the voltage rating of a low-capacitance probe. Such points are generally marked "Do not measure" in the receiver service data.

7

COLOR TV
TROUBLESHOOTING
WITH THE
OSCILLOSCOPE

7.1 GENERAL CONSIDERATIONS

A color-TV receiver contains all of the sections employed in a black-and-white receiver, plus the color-circuit sections depicted in Fig. 7-1. Oscilloscope tests are made in all of these sections in order to localize operating troubles. As previously explained for black-and-white receivers, basic oscilloscope tests involve waveform analysis, peak-to-peak voltage measurements, and checks of frequency-response curves. However, somewhat different types of waveforms are encountered in chroma (color) circuitry. As an illustration, chroma-demodulator operation is based on phase response, and demodulation action is often checked by means of specialized Lissajous figures called *vectorgrams*. These waveforms are explained in detail subsequently.

Some basic characteristics of the composite color video signal are shown in Fig. 7-2. A color burst appears on the back porch of the horizontal sync pulse. This color burst consists of eight or nine cycles of 3.58-MHz sine-wave (color-subcarrier) signal. The color burst normally has

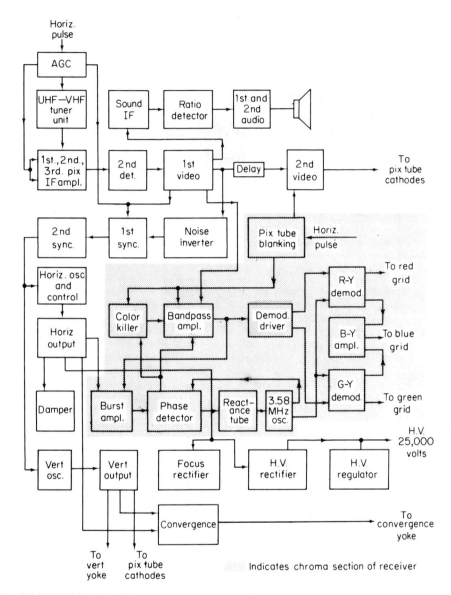

Fig. 7-1. Block diagram of a color-TV receiver.

the same amplitude as the horizontal sync tip. Note the makeup of the yellow-bar signal in Fig. 7-2. It consists of a 3.58-MHz sine-wave component centered on a brightness (Y) component. There is a fundamental distinction in phase between the color burst and the yellow-signal chroma

171

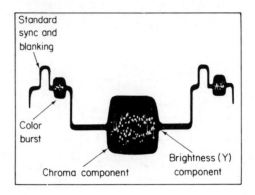

Fig. 7-2. Basic color-TV waveform, showing the color burst and a yellow bar signal.

component, as seen in Fig. 7-3. That is, the burst signal has a phase angle of zero degrees, whereas the yellow signal has a phase angle of 12.5 deg. Vectorgrams are utilized to measure chroma phase angles.

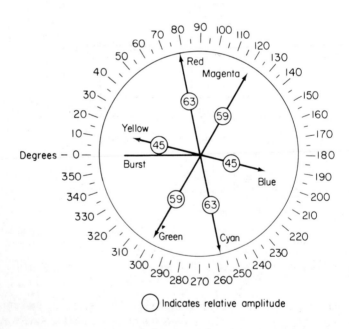

Fig. 7-3. Chroma phase angles for the basic colors, with respect to the color-burst phase.

172

7.2 DELAY LINE TESTS

A delay line imposes a time delay of approximately 1 microsecond on the Y signal. Although this delay time can be measured directly with a triggered-sweep oscilloscope, most technicians restrict their tests to input and output waveform checks, as illustrated in Fig. 7-4. Note that a delay line produces some minor distortion of a horizontal sync pulse. However, the amount of distortion shown in the example is within normal operating tolerance. Note also that the amplitude of the output waveform should be practically the same as the amplitude of the input waveform. The correct peak-to-peak voltage value will be specified in the receiver service data.

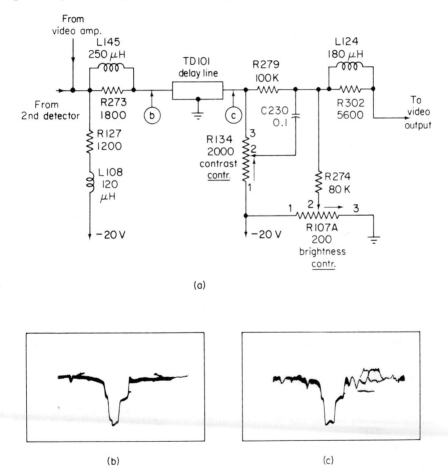

Fig. 7-4. Delay-line waveform checkout. (a) Circuitry. (b) Input waveform. (c) Output waveform.

When it is desired to measure the delay time, the test setup depicted in Fig. 7-5 is utilized. As the sync pulse enters the delay line, it starts the horizontal sweep via the external-trigger function of the oscilloscope. Then, as the pulse emerges from the delay line, it is displayed on the CRT screen. By counting the number of centimeters covered by the delay-time interval, and noting the sweep-speed setting of the oscilloscope, we can obtain the delay time of the line. Most delay-line defects are caused by mechanical damage, or poor connections.

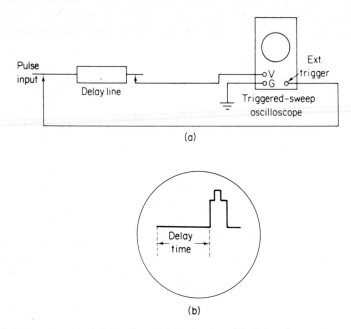

Fig. 7-5. Measurement of delay time. (a) Test setup. (b) Delay-time pattern.

7.3 BURST AMPLIFIER TESTS

A burst amplifier functions as a color-sync separator. The color burst is gated out of the composite color signal by a keying pulse, as depicted in Fig. 7-6. An important requirement is for the keying pulse to be correctly timed with respect to the burst. Timing of the keying pulse is determined in part by the setting of the horizontal-hold control. Accordingly, the burst amplifier should be checked with the horizontal-hold control set to the midpoint of its range. A test setup for checking the burst amplifier and the normal waveforms is shown in Fig. 7-7. Peak-to-peak voltages are specified in the receiver service data. In case of distorted or weak wave-

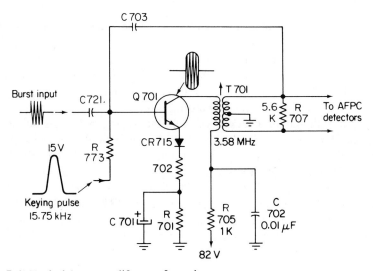

Fig. 7-6. Typical burst-amplifier configuration.

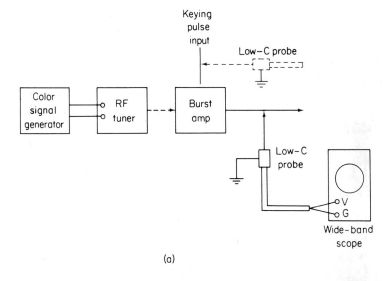

(a)

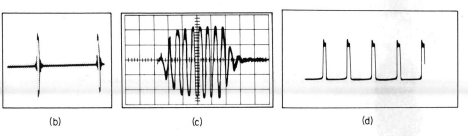

(b)   (c)   (d)

Fig. 7-7. Testing the burst-amplifier section. (a) Test setup. (b) Normal output waveform. (c) Expanded output waveform. (d) Normal keying pulse.

175

forms in the burst-amplifier section, the amplitude of the burst input waveform should be checked. If this waveform has subnormal amplitude, the trouble will be found in a prior receiver section.

In case the burst input waveform is normal, the amplitude of the keying pulse should be checked next. If the keying pulse has subnormal amplitude, the trouble will be found in the keying-pulse branch. On the other hand, if the keying pulse is normal, there is a defect in the burst-amplifier circuitry. With reference to Fig. 7-6, C701 may have lost capacitance, or be open-circuited. C721, C702, and C703 are also likely suspects. In case there are no capacitor defects, CR715 should be checked for front-to-back ratio. Collector-junction leakage in Q701 can also cause a weak output waveform. Although least likely, it is possible for T701 to develop a short- or open-circuited winding.

7.4 CHROMA BANDPASS AMPLIFIER TESTS

Oscilloscope tests in the bandpass-amplifier section consist of waveform and frequency-response checks. Waveform tests are made at the input and output of the bandpass amplifier, as depicted in Fig. 7-8. A keyed-rainbow signal is generally applied to the antenna-input terminals of the receiver. In turn, the video amplifier applies the waveform illustrated

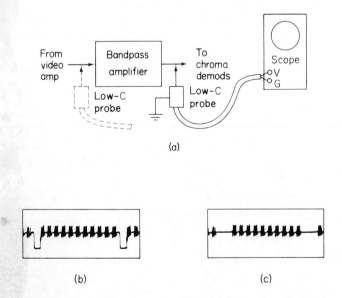

Fig. 7-8. Waveform check of bandpass amplifier. (a) Test setup. (b) Input and output waveforms.

in Fig. 7-8(b) to the input of the bandpass amplifier. Note that a keyed-rainbow signal consists of a horizontal sync pulse followed by 11 "bursts." Each burst is a train of 3.56-MHz sine waves (3.579545–0.015750 MHz). The first burst following the sync pulse operates as a color burst for the color sync system. This burst and the horizontal sync pulse are gated out in many receivers, as seen in Fig. 7-8(c). In other receivers, only the horizontal sync pulse is eliminated, and the color burst passes through the bandpass amplifier.

When troubleshooting the bandpass amplifier, the essential points in waveform analysis are to check the input and output peak-to-peak voltages, and to determine whether the gating action is occurring normally. If the input waveform has subnormal amplitude, the trouble will be found in the preceding circuit sections. On the other hand, if the input waveform is normal and the output waveform is weak, the trouble is located in the bandpass-amplifier section. Faulty capacitors are the most common troublemakers. If the color burst appears in the output waveform, although the receiver service data indicate that it is gated out, it is most probable that the gating pulse is attenuated or missing. In turn, the gating-pulse circuit should be checked for defective components.

Alignment of a bandpass amplifier requires a video sweep and marker generator, because the circuits have appreciable bandwidth. By way of comparison, a burst amplifier such as that shown in Fig. 7-6 is merely peaked for maximum output at 3.58 MHz. On the other hand, a bandpass amplifier generally has a rising response to compensate for the falling response of the video-IF curve, as depicted in Fig. 7-9. This compensation provides a uniform overall response through the receiver to the output of the bandpass amplifier. A simple alignment test setup is shown in Fig. 7-10. The tuned circuits in the bandpass-amplifier section are adjusted to obtain a frequency-response curve as specified in the receiver service data. The top of the response curve usually extends from 3.1 to 4.1 MHz.

A more sophisticated check of system frequency response is made with the video sweep modulation (VSM) method, as depicted in Fig. 7-11. In other words, the system response to the output of the bandpass amplifier includes the RF, IF, first-video, and bandpass-amplifier tuned circuits. Because of operating tolerances in each of these sections, a compensating alignment adjustment is often desirable in the bandpass-amplifier section. To make a VSM alignment check, a video sweep signal must be encoded into the VHF picture-carrier signal. This is done by externally modulating a VHF signal generator (marker generator) with a video-frequency sweep signal. Thereby, a system response curve is obtained at the output of the bandpass amplifier, as exemplified in Fig. 7-11. The bandpass amplifier is aligned for uniform response from 3.0 to 4.1 MHz in this example.

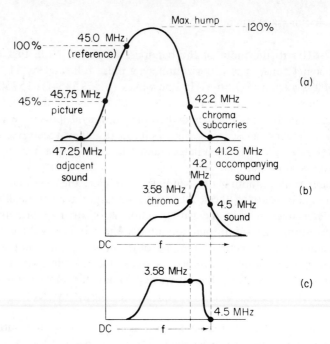

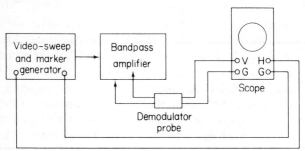

Fig. 7-9. Bandpass amplifier aligned to compensate for falling response in the IF section. (a) IF response curve. (b) Bandpass-amplifier response curve. (c) Overall IF and bandpass response curve.

Phased 60-Hz sine-wave horizontal deflection voltage

(a)

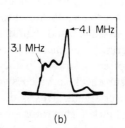

(b)

Fig. 7-10. Bandpass-amplifier alignment test setup. (a) Instrument connections. (b) Typical response curve.

178

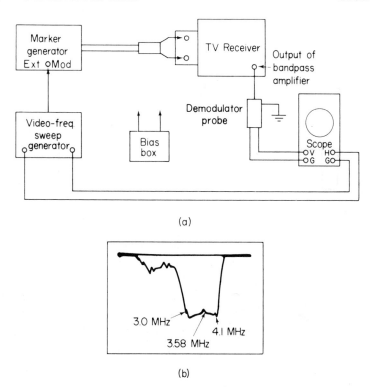

(a)

(b)

Fig. 7-11. Video sweep modulation frequency-response test. (a) Instrument connections. (b) Typical VSM response curve.

7.5 CHROMA DEMODULATOR TESTS

Chroma demodulation is the process of decoding the complete color signal. Figure 7-12 shows how the chroma signal is encoded at the color-TV transmitter (or in a color-bar generator). Encoding consists of combining the 3.58-MHz chroma signal with the Y or brightness signal. It also involves a specific phasing of each chroma signal. That is, certain phases with reference to burst correspond to the primary colors red, green, and blue, and to the complementary colors cyan, magenta, and yellow. The chroma phases employed in this encoding process are indicated in Fig. 7-13. In the color-TV receiver, the decoding or chroma demodulation process takes place in reverse, with the result that the primary and complementary color signals are recovered from the complete color signal.

Chroma demodulation involves both phase and amplitude detection, as depicted in Fig. 7-14. Note that the basic chroma-demodulator circuit is similar to a ratio-detector configuration. Thereby, phase detection is

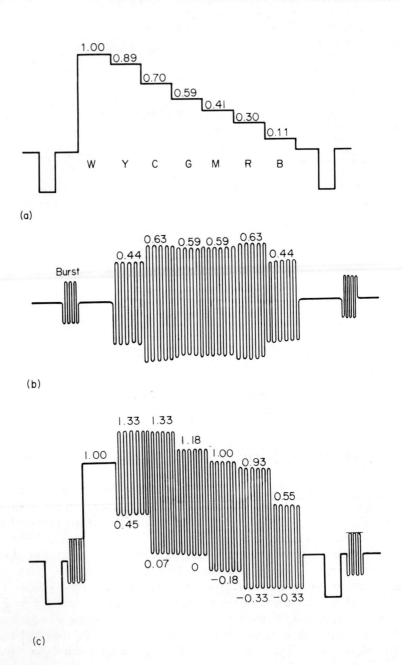

Fig. 7-12. Encoding of the complete color signal. (a) Y or brightness component. (b) Chroma component. (c) Complete color signal.

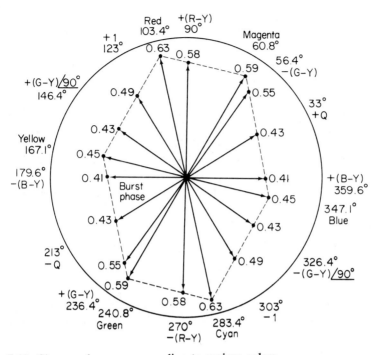

Fig. 7-13. Chroma phases corresponding to various colors.

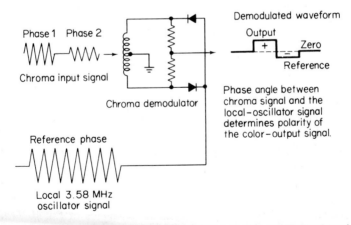

Fig. 7-14. Chroma demodulation involves both phase and amplitude detection.

obtained. On the other hand, unlike a ratio detector, a chroma demodulator does not employ a stabilizing capacitor. In turn, the amplitude of the output signal is proportional to the amplitude of the input signal.

Note that the Y signal is rejected by the bandpass amplifier, and that only the 3.58-MHz chroma signal is applied to the chroma demodulator. After the chroma signal has been demodulated, the Y signal is added to the demodulated waveform. This addition, or matrixing, may take place in the color picture tube, or in a circuit between the chroma demodulator and the picture tube.

In practice, three chroma demodulators are usually employed. These are called R-Y, B-Y, and G-Y demodulators. They decode the chroma components of the red, blue, and green color signals. As an illustration, if the chroma demodulator depicted in Fig. 7-14 were an R-Y demodulator, the reference phase would be set at 90 deg, as seen in Fig. 7-13. Or, if it were a B-Y demodulator, the reference phase would be 359.6 deg. Again, if it were a G-Y demodulator, the reference phase would be 236.4 deg. Chroma-demodulator operation is usually checked with a keyed-rainbow signal, as shown in Fig. 7-15. Note that the R-Y waveform normally crosses over the zero axis on the sixth pulse. The B-Y waveform normally crosses over on the third and ninth pulses. The G-Y waveform normally crosses over on the first and seventh pulses.

If a chroma waveform crosses over the zero axis incorrectly, there is a demodulator phasing error present. This trouble will usually be found in the 3.58-MHz reference circuitry between the subcarrier oscillator and the affected chroma demodulator. Leaky capacitors are common trouble-makers. A chroma waveform is also checked for peak-to-peak voltage. Subnormal amplitude is generally caused by a defective component in the demodulator circuit. Note that if a demodulator diode develops a poor front-to-back ratio or becomes open-circuited, the output waveform amplitude will be subnormal and a crossover error will also be observed. Note also that if the reference voltage from the subcarrier oscillator is stopped, as by a short-circuited capacitor, there will be no output from the chroma demodulator.

The baseline or zero level in the Fig. 7-15 waveforms is virtually straight. This characteristic of the display indicates that the chroma-demodulator circuit has normal bandwidth (approximately 0.5 MHz). On the other hand, when a baseline has excessive curvature, it is indicated that the chroma-demodulator circuit has subnormal bandwidth. As an illustration, the G-Y waveform in Fig. 7-16 displays excessive curvature. Although there is a little curvature in the baselines of the R-Y and B-Y waveforms, this is within normal tolerance. To troubleshoot the G-Y demodulator circuit, fixed capacitors and peaking coils are tested. If a peaking coil has become short-circuited, the demodulator bandwidth will be reduced and color reproduction impaired.

Vectorgrams are also used to check chroma-demodulator action. A

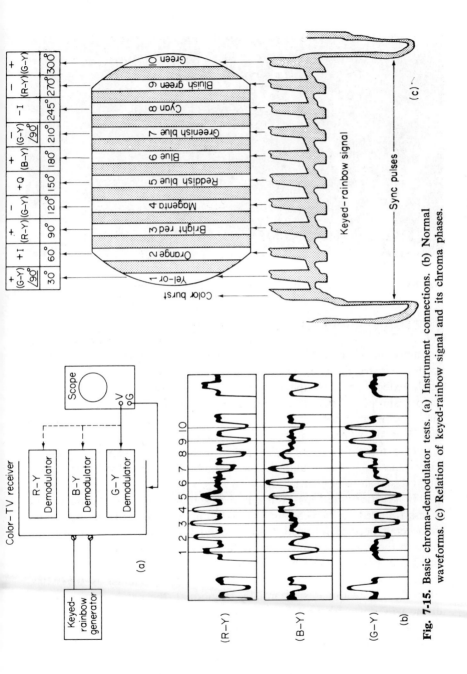

Fig. 7-15. Basic chroma-demodulator tests. (a) Instrument connections. (b) Normal waveforms. (c) Relation of keyed-rainbow signal and its chroma phases.

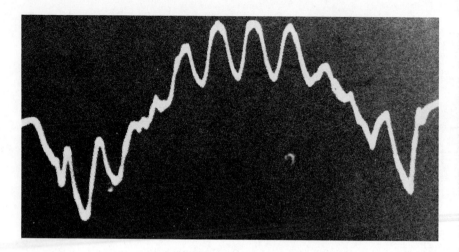

Fig. 7-16. The G-Y waveform has excessive baseline curvature.

vectorgram is a particular type of Lissajous figure obtained as shown in Fig. 7-17. Although ideal vectorgrams are not obtained in practice, it is instructive to observe the features of the ideal waveform. Note that the petals have straight sides and tops that are segments of a circle. Also, the petals extend down to the center of the ideal vectorgram. In practice, vectorgram petals depart considerably in various ways from the ideal shape, as exemplified in Fig. 7-18. Note also that an ideal vectorgram has 12 petals, whereas an actual vectorgram has 10 petals. This difference is due to the gating or blanking pulse in the chroma section, which removes the color burst and the horizontal sync pulse from the keyed-rainbow signal, as shown in Fig. 7-18.

In practice, the tops of the keyed-rainbow waveform become more or less rounded, as depicted in Fig. 7-19. In turn, the tops of the vectorgram petals become rounded. Corner rounding is caused by limiting bandwidth in the chroma circuits, just as the corners of a square wave become rounded in passage through an amplifier with limited bandwidth. As noted previously, the normal bandwidth of the chroma channel is 0.5 MHz. When the bandwidth is substantially subnormal, the petals do not extend all the way down to the center of the vectorgram, as exemplified in Fig. 7-20. The reason for this type of display is seen from Fig. 7-16, in which the G-Y keyed-rainbow waveform has excessive baseline curvature.

Vectorgram petals correspond to the hues noted in Fig. 7-21, which are displayed on the screen of a color picture tube by a keyed-rainbow

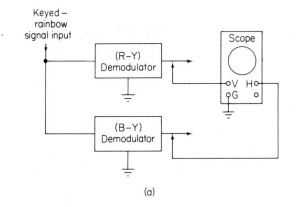

(a)

R-Y signal applied
to vertical deflection plates

B-Y signal applied to
horizontal deflection
plates

(b)

Fig. 7-17. Development of an ideal vectorgram pattern. (a) Instrument connections. (b) Keyed-rainbow signals and vectorgram.

185

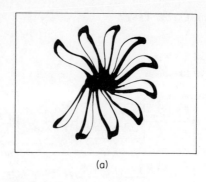

(a)

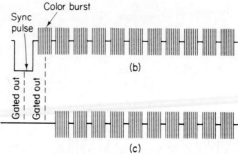

(b)

(c)

Fig. 7-18. Typical 10-petal vectorgram displayed in practice. (a) Vectorscope display. (b) Complete keyed-rainbow signal. (c) Sync-pulse and color-burst components gated out.

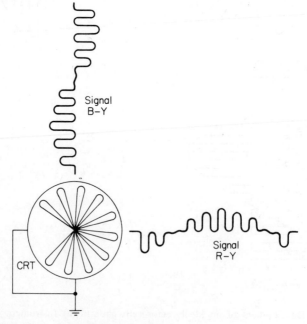

Fig. 7-19. Formation of rounded petal tops.

186

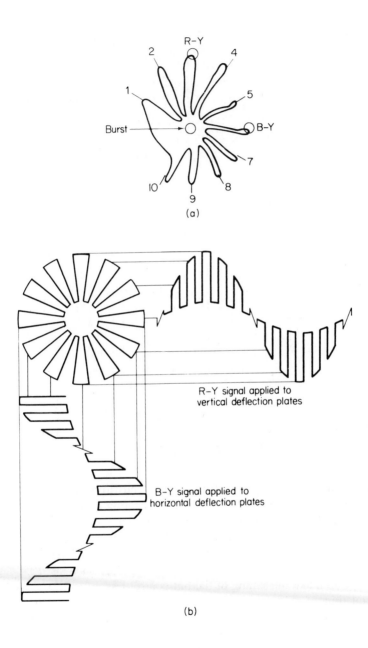

Fig. 7-20. Vectorgram display with central open area. (a) Typical pattern. (b) How central open area is developed.

187

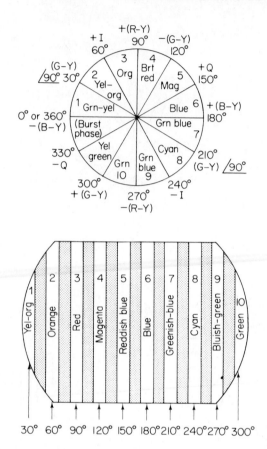

Fig. 7-21. Vectorgram petal positions correspond to hues displayed on the picture-tube screen by a keyed-rainbow signal.

signal. A circular vectorgram outline corresponds to a demodulation angle of 90 deg. As an example, R-Y/B-Y demodulation occurs along axes that have a phase difference of 90 deg, and a circular vectorgram is produced. On the other hand, various color-TV receivers have demodulation angles other than 90 deg, and these receivers produce elliptical vectorgram patterns. A demodulation angle greater than 90 deg, such as 105 deg, is employed to make the setting of the tint control less critical for satisfactory reproduction of flesh tones (hues in the orange range). Although this mode of demodulation introduces some distortion of hues, the trade-off is deemed justified by various manufacturers.

Chroma demodulators that operate with phase angles other than 90 deg are generally called XZ demodulators. Figure 7-22 depicts a 105-deg

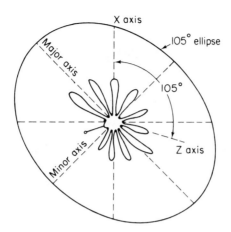

Fig. 7-22. A vectorgram produced by XZ demodulators operating with a 105-deg phase angle.

ellipse, and a vectorgram produced by XZ demodulators operating at a 105-deg phase angle. Note that this pattern aspect is obtained when the vertical- and horizontal-amplifiers of the vectorscope are set for equal gain. This results in a major axis that inclines at an angle of 45 deg with respect to the horizontal axis. Some chroma demodulators operate at other phase angles, and the receiver service data should be consulted in any case. Note that if the demodulation angle is not specified otherwise, it may be determined by inspection of the demodulator output waveforms, as shown in Fig. 7-15. Thus, the R-Y waveform crosses at the sixth pulse, and the B-Y waveform crosses at the third pulse. Since each pulse corresponds to a 30-deg phase interval, the R-Y and B-Y demodulators operate at a 90-deg phase angle.

A complete vectorgram checkout of the chroma system requires display of the G-Y signal also. For example, the output signals from the B-Y and G-Y demodulators may be used as depicted in Fig. 7-23. There is a phase difference of 120 deg between the B-Y and G-Y signals. Note that some XZ demodulators operate at a 120-deg phase angle, also. Although it is possible to check G-Y demodulator action by displaying an R-Y/G-Y vectorgram, this is less desirable because the phase angle involved is 150 deg, as seen in Fig. 7-15. In turn, the minor axis of the vectorgram is quite short, and it is comparatively difficult to evaluate the pattern. Therefore, it is preferred to utilize a B-Y/G-Y vectorgram.

Overloading in a chroma channel shows up in two ways. If the peak of a chroma signal is being clipped, a "flat" appears in the vectorgram

189

outline, as shown in Fig. 7-24. Overloading is usually caused by incorrect bias of a transistor in the bandpass amplifier. Sometimes, an inexperienced technician will turn the color control too high, and thereby produce overloading of an otherwise normally operating chroma system. When an

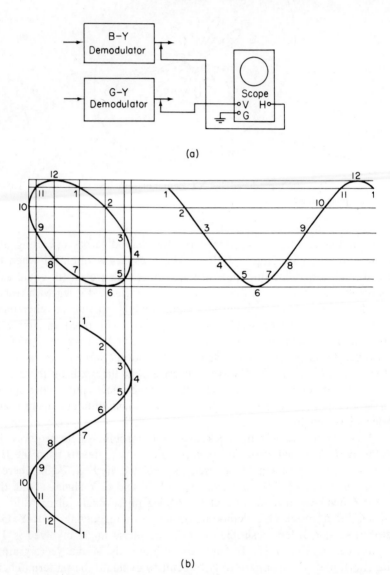

(a)

(b)

Fig. 7-23. Display of a B-Y/G-Y vectorgram. (a) Instrument connections. (b) Development of a 120-deg ellipse.

overload condition causes peak compression instead of clipping, the vectorgram becomes egg-shaped, as exemplified in Fig. 7-25. Collector-junction leakage in a transistor can cause this symptom, for example. Again, a demodulator diode might have a poor front-to-back ratio.

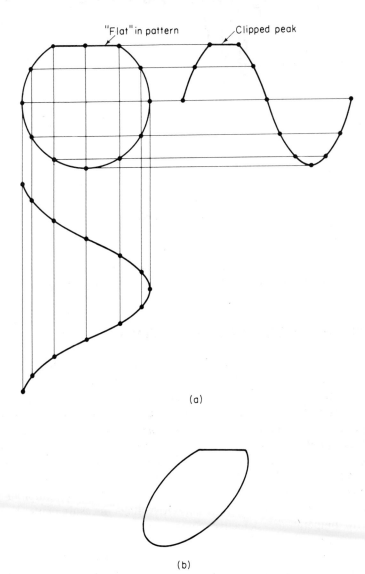

(a)

(b)

Fig. 7-24. Overload indication in vectorgram outline. (a) Development of "flat" in a circular pattern. (b) Appearance of "flat" in an elliptical pattern.

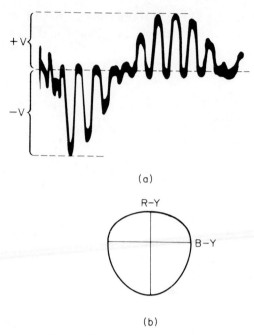

Fig. 7-25. Peak compression produces distorted waveforms. (a) Positive-peak compression of a keyed-rainbow waveform. (b) Egg-shaped vectorgram outline.

7.6 OTHER COLOR-TV WAVEFORMS

In addition to the waveforms that have been discussed, other waveforms can be checked in the color-sync section and the convergence section. However, the principles that are involved are basically the same as explained previously. In each case, the waveshape and its peak-to-peak voltage are compared with the specifications in the receiver service data, with due allowance for operating tolerances. It follows from foregoing examples that waveform analysis occasionally serves to pinpoint a defective component, such as an open coupling capacitor. In most situations, however, waveform analysis can only localize a trouble symptom to a particular stage or circuit. In turn, DC-voltage measurements must be made to pinpoint the defective component. Sometimes, resistance measurements are also required to supplement voltage measurements.

Some receivers have an automatic tint control (ATC) that serves to make the setting of the tint control less critical for satisfactory reproduction of flesh tones. When checking the chroma demodulators, it is essential to turn off the ATC control. Otherwise, the waveforms and vectorgrams will be nonstandard and will not correspond to service-data specifications. An ATC network functions to change the reproduced colors in the vicinity of orange, such as yellow and red, into orange. Thus, a trade-off is involved in which more or less color distortion is accepted in return for constancy of flesh tones when one is switching stations. If the receiver service data happens to specify operating waveforms for the ATC function, the ATC control may be turned on for an oscilloscope check.

7.7 VECTORGRAM APPLICATION IN COLOR-TV TRANSMISSION

Color-TV transmitters employ NTSC color-bar signals, instead of keyed-rainbow signals, to check system operation. In turn, the vectorgram produced by an NTSC color-bar signal differs considerably from the vectorgrams that have been noted previously. Figure 7-26 shows a complete color-bar signal produced by an NTSC generator, and its chroma component. An NTSC vectorgram shows the phase relations of these chroma components, and their relative amplitudes. Normal NTSC chroma signal phases and amplitudes are depicted in Fig. 7-27. To check these phases and amplitudes, the transmitter is energized by an NTSC color-bar pattern.

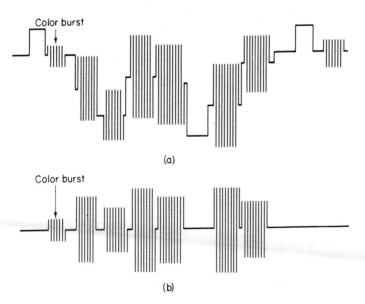

(a)

(b)

Fig. 7-26. An NTSC color-bar signal waveform. (a) Complete signal. (b) Chroma component.

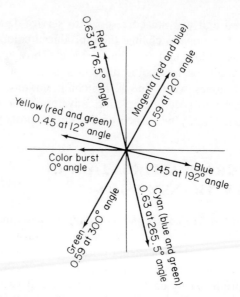

Fig. 7-27. Normal NTSC chroma signal phases and amplitudes.

A sample of the transmitter output signal is then applied to a pair of R-Y and B-Y demodulators, and the demodulator outputs are applied to a vectorscope in the usual manner. However, a vectorscope graticule for use with an NTSC color-bar signal has the form depicted in Fig. 7-28. The small circles indicate permissible tolerances on the phases and amplitudes of the chroma signals. The displayed vectorgram is developed as

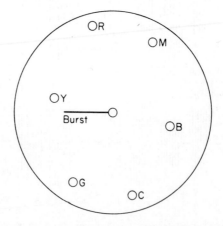

Fig. 7-28. Vectorscope graticule for use with an NTSC color-bar signal.

shown in Fig. 7-29. It appears essentially as a pattern of bright dots (beam-resting positions), although the transition lines from one dot to the next may appear faintly in the pattern.

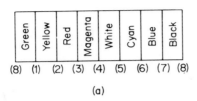

(a)

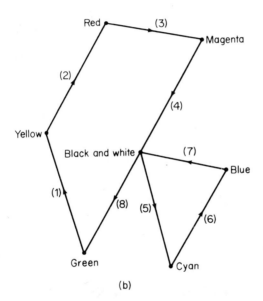

(b)

Fig. 7-29. Development of an NTSC vectorgram pattern. (a) Color-bar sequence. (b) Vectorgram.

8

OSCILLOSCOPE
MAINTENANCE

8.1 GENERAL CONSIDERATIONS

Trouble symptoms in oscilloscope operation include dark screen, inability
to center the pattern, no vertical deflection, no horizontal deflection, poor
focus, no sync action, no retrace blanking, overheating and/or smoking,
drift in pattern position, nonlinearity, motorboating, poor sensitivity, impaired frequency response, and various intermittent conditions. Sometimes
an apparent trouble symptom is actually caused by incorrect control set-
ting(s). Therefore, this possibility should be investigated at the outset. In
case of a dark screen and dark pilot light, check also to determine if there
is AC power from the outlet. Before proceeding to localize circuit trouble
in an oscilloscope, the block diagram should be reviewed. Figure 8-1 shows
a typical block diagram with associated operating controls indicated.

Warning

The cathode-ray tube in an oscilloscope operates at a high voltage, which can cause injury or death if accidentally contacted. Only experienced technicians should be per-

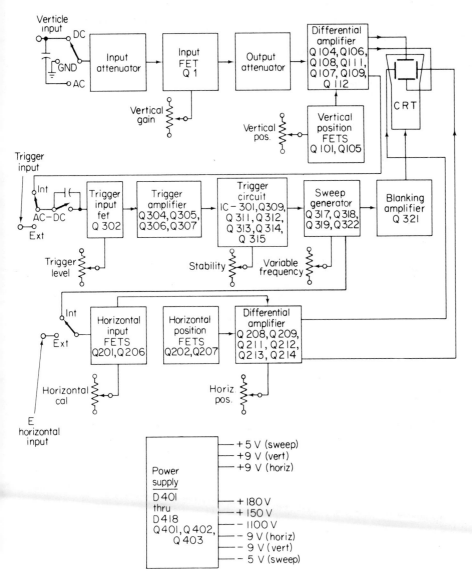

Fig. 8-1. Block diagram of a typical oscilloscope. (*Courtesy of* Heath Co.)

197

mitted to operate an oscilloscope with its case removed.
The power supply in an oscilloscope can be deceptive,
since the heater and cathode of the CRT are operated at
1000 volts or more above ground. Even when an oscilloscope is turned off, the filter capacitors may hold a deadly
charge for a long time if the high-voltage bleeder resistor
opens up. Therefore, always discharge the filter capacitors
before starting to troubleshoot oscilloscope circuitry. Remember also that circuit defects can cause dangerously
high voltages to appear at unexpected points in oscilloscope networks.

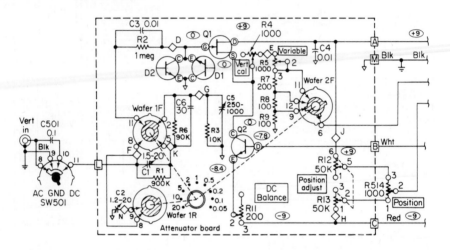

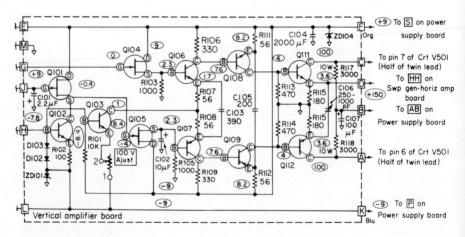

Fig. 8-2. Schematic diagram of a typical vertical amplifier. (*Courtesy of* Heath Co.)

It is sometimes helpful to observe the action of the operating controls. As an illustration, the vertical-position control in Fig. 8-2 is connected at the input of the vertical differential amplifier, and it affects the DC voltages in each of the following stages. Suppose that a trace is visible on the CRT screen, but there is no response to variation of the position-control setting. The power-supply voltages should be verified first. Then it is advisable to check the collector voltages of transistors Q111 and Q112 in Fig. 8-2. These voltages normally vary as the vertical-position control is turned. If these voltages vary, the trouble will be found in the CRT circuit. On the other hand, if the voltages do not vary, the trouble is in Q111, Q112, or a preceding stage. Thus, the next check is made at the collectors of Q108 and Q109. By checking back toward the vertical-position control, the trouble can be localized to a particular stage.

8.2 PINPOINTING COMPONENT DEFECTS

Because most of the circuits are DC-coupled in configurations such as Fig. 8-2, the possibility of listing "cause and effect" types of component defects is greatly restricted. That is, there is considerable interaction among the various circuit branches. As an example, a saturated transistor on one side of a differential amplifier may appear as a trouble condition on the other side of the amplifier. DC-voltage readings can clear an area or a component from suspicion, but they have limited ability in pinpointing a defective component. Thus, if the vertical amplifier in Fig. 8-2 is "dead," and the indicated DC-voltage values are measured (within a tolerance of ±20 per cent), the active portion of the vertical amplifier would be cleared from suspicion. In turn, the attenuator network would be checked.

Again, many of the indicated DC-voltage values in Fig. 8-2 might be incorrect, showing that there is a defective component in the active portion of the vertical amplifier. In this situation, the best procedure is to make signal-tracing or signal-substitution tests. That is, an input signal can be applied to the vertical amplifier, and the signal can be traced step-by-step through the amplifier with another oscilloscope. In turn, the point at which the signal stops localizes the defective stage. Sometimes a signal-tracing test will pinpoint a defective component. For example, if Q1 is short-circuited, the signal waveform will be found at the gate, but not at the source terminal. A signal-substitution test will provide the same troubleshooting information. An audio oscillator is a convenient signal source. A 0.1-μf capacitor should be connected in series with the "hot" lead from the audio oscillator, to avoid drain-off of DC voltages while the signal is injected.

When a component such as a resistor or a capacitor is defective, the fault can often be pinpointed by checking circuit resistance values with a hi-lo ohmmeter. A hi-lo ohmmeter has a low-voltage resistance-measuring function that applies less than 0.1 volt between the terminals under test. In turn, transistor and diode junctions are not "turned on" by the test voltage, and semiconductor devices appear as open circuits to the ohmmeter. With reference to Fig. 8-2, the resistance values of R117 and R118 can be measured in-circuit with a hi-lo ohmmeter because the collector junctions of Q111 and Q112 will not conduct during the test. Similarly, the resistance values of R115 and R116 can be measured, because the base-emitter junctions of Q111 and Q112 will not conduct during the test. Again, the resistance values of R113 and R114 can be measured.

A hi-lo ohmmeter also permits in-circuit tests for "shorted" capacitors such as C103 in Fig. 8-2. In other words, the resistance between the emitter terminals of Q106 and Q107 will read 112 ohms ±20 per cent if C103 is not short-circuited. On the other hand, a reading substantially less than 88 ohms indicates serious leakage or a short circuit in C103. Of course, a hi-lo ohmmeter cannot indicate a capacitor that is open-circuited. If the stage gain is low and it is suspected that C103 is open-circuited, a "bridging" test is advisable. This test is made by temporarily shunting a known good capacitor across C103, to determine whether the normal stage gain is restored. "Bridging" tests can also be made of electrolytic capacitors, such as C104. Operating controls such as R5 and R514 may become worn and "noisy" eventually. A hi-lo ohmmeter is useful to check the resistance variation of a suspected control as it is turned through its range.

8.3 VERTICAL ATTENUATOR
TROUBLESHOOTING

A vertical step attenuator must be adjusted for correct compensation in order to avoid distortion of complex waveforms. With reference to Fig. 8-2, the portion of the attenuator on the gate side of Q1 is compensated, whereas the portion on the source side is not compensated. Compensation is required on the gate side because the input impedance of the vertical channel is high (1 megohm). On the other hand, the impedance of the source circuit is low (less than 2000 ohms). In turn, stray capacitances do not affect the operation of the source circuit appreciably, and compensation is not required. High-frequency compensation is obtained by adjustment of trimmer capacitors C1, C2, and C5.

Figure 8-3 shows the basic configuration of the vertical step attenuator. It is essentially a voltage-divider arrangement with three steps, *A, B,* and *C.* When trimmer capacitors C1, C2, and C5 are correctly adjusted, complex waveforms are passed without distortion into output terminals *A, B,* and *C.* The trimmers can be adjusted to good advantage on the basis of square-wave reproduction, as depicted in Fig. 8-4. A square-wave repetition rate in the range from 1 kHz to 10 kHz is suitable. Note that capacitor C2 in Fig. 8-3 is a switching-capacitance compensator. In other words, when the vertical-attenuator switch is changed from *A* to *B,* the stray capacitance of the switch is transferred from *A* to *B.* To compensate for this capacitance change, C2 is automatically switched into the circuit.

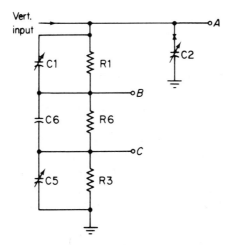

Fig. 8-3. Vertical step-attenuator configuration.

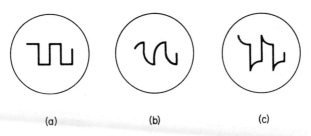

(a)   (b)   (c)

Fig. 8-4. Step-attenuator adjustments. (a) Correct trimmer capacitance. (b) Insufficient capacitance. (c) Excessive capacitance.

After the vertical step attenuator has been properly compensated, the low-capacitance probe for the oscilloscope should be checked for compensation. The probe (Fig. 8-5) is connected to the vertical-input terminal of the oscilloscope, and trimmer capacitor C_T is adjusted for optimum square-wave reproduction. A low-capacitance probe should be designed for the oscilloscope with which it is used, in order to obtain a standard attenuation ratio of 10 to 1. The attenuation ratio is established by the value of R in Fig. 8-5.

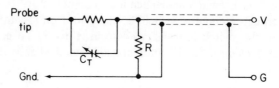

Fig. 8-5. A low-capacitance probe configuration.

8.4 VERTICAL CALIBRATION

After the vertical step attenuator has been properly compensated, vertical calibration may be checked and adjusted if necessary. A square-wave generator is a convenient source of calibrating voltage, and its value in peak-to-peak volts can be measured with a TVM, as depicted in Fig. 8-6. With reference to Fig. 8-2, the variable attenuator R5 must be turned to its zero-resistance position. The vertical step attenuator may then be set to moderate sensitivity position, such as 1 volt per centimeter. In turn, the output from the square-wave generator (Fig. 8-6) is adjusted to obtain a reading of 1 volt p-p on the TVM. If the waveform displayed on the CRT screen occupies more or less than 1 centimeter vertically, the vertical-calibration control (R4 in Fig. 8-2) is adjusted as required.

When the oscilloscope is in calibration on any given step of the vertical attenuator, it will normally be in calibration on every step also. However, in case the attenuator is found to be off calibration on one or more steps, look for a leaky capacitor or off-value resistor in the step-attenuator circuit. For example, leakage in C6 (Fig. 8-2) will upset the calibration on various steps. If R1, R6, R3, R7, R8, or R9 is off-value,

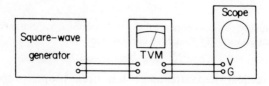

Fig. 8-6. Calibration test setup.

the calibration will be upset on one or more steps. Note also that the attenuation factor of the low-capacitance probe should be checked at intervals. In normal operation, the probe introduces a 10-to-1 attenuation. Thus, it changes the 1 volt per centimeter step of the vertical attenuator to a 10 volts per centimeter step. If the probe factor is in error, look for an off-value resistor in the probe circuit.

8.5 POSITION AND DC BALANCE ADJUSTMENTS

If the position of the trace tends to shift as the variable gain control is turned, the two sides of the differential amplifier are not balanced with respect to the DC voltage distribution. In most cases, only an adjustment of the DC balance control is required. With reference to Fig. 8-2, R11 is carefully set to the point where the pattern remains in the same place on the CRT screen as R5 is varied. This setting corresponds to zero volts on the source of Q1. Next, the voltages on the collectors of Q111 and Q112 should be measured. If they are not equal, set R514 to the center of its range, and adjust R12 and R13 to make the collector voltages equal. In this example, the collector voltages should be 100 volts at this time. If not, R104 is adjusted as required. Note that there is some interaction between R104 and R12-R13.

In case the foregoing adjustments cannot be made satisfactorily, it is most likely that the differential amplifier is unbalanced owing to mismatched transistors. As an illustration, if Q108 develops collector-junction leakage, the DC voltage distribution will be upset. Because DC coupling is employed in this example, it is not always possible to pinpoint a faulty transistor by analysis of the DC voltage distribution. However, a pair-turnoff test can be made in this situation which provides useful data. For example, if Q108 has collector-junction leakage, this fact will show up as follows:

1. When the base and emitter of Q111 are connected together with a jumper lead, and the base and emitter of Q112 are connected together with a jumper lead, the collector voltages of Q111 and Q112 become equal. Therefore, the trouble is not in Q111 or Q112, because both of the transistors cut off normally.

2. Remove the jumpers from Q111 and Q112, and connect the base and emitter of Q108 together, and connect the base and emitter of Q109 together. The collector voltages of Q111 and Q112 will now be unequal, because Q108 does not cut off normally. The

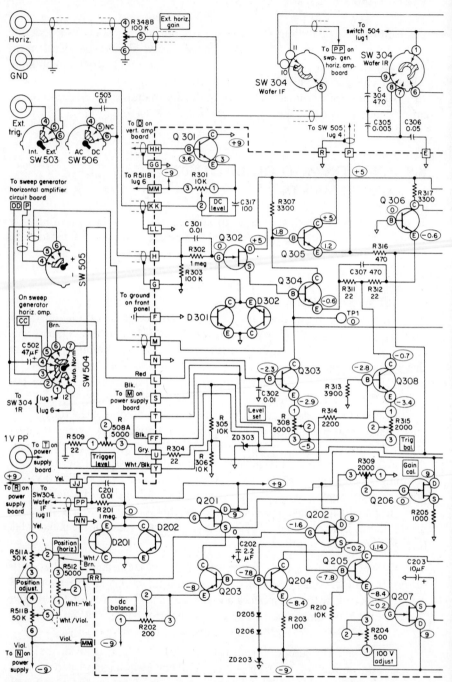

Fig. 8-7. A triggered-sweep time base and horizontal-amplifier configuration. (*Courtesy of* Heath Co.)

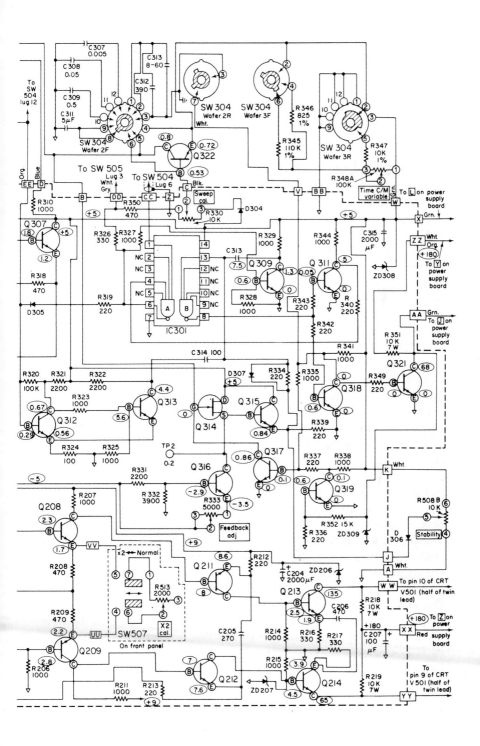

205

collector voltage of Q108 is a bit lower than the collector voltage of Q109 in this situation, which pinpoints the trouble to Q108.

The same general troubleshooting principles apply to horizontal amplifiers, as exemplified in Fig. 8-7.

8.6 TIME-BASE CALIBRATION

A triggered-sweep time base and horizontal-amplifier configuration is shown in Fig. 8-7. To calibrate the time base, a square-wave pattern is displayed on the CRT screen, and the gain-calibration control R309 is set as required to obtain a correct time/cm display. For example, suppose that the sweep speed is set to 100 μs/cm, and the square-wave generator is set to a repetition rate of 1250 Hz. Then, the gain-calibration control should be adjusted to make one cycle of the square wave occupy 8 centimeters horizontally, as depicted in Fig. 8-8. Note that the accuracy of calibration depends on the accuracy of the square-wave generator.

After the time base has been calibrated on one sweep-speed setting, it will normally be in calibration on the other sweep-speed settings. However, if an error is found on other settings, there is a defective component in the time-base circuit. Note that when SW507 is set to its X2 position, the time base may seem to be out of calibration because R513 is out of adjustment. With reference to Fig. 8-5, R513 should be adjusted so that in the X2 position of SW507, a half cycle of the square wave occupies 8 cm instead of 4 cm. Note that the calibration adjustments will be incorrect throughout if the operator forgets to turn the variable time/cm control R348A off (Fig. 8-7).

Troubleshooting a time base starts with observation of operating control responses. For example, if operation is incorrect on one particular switch position, it is indicated that the trouble will be found in the associated switching circuit. In case the time base is "dead," systematic DC-voltage measurements are made to localize the general trouble area. The normal operating voltages exemplified in Fig. 8-7 are measured when the oscilloscope controls are set to the following positions. Trigger switches SW503, SW504, SW505, and SW506 are set to their Int., Auto, —, and AC positions, respectively. The time/cm switch SW304 is set to its fully clockwise (Ext. In.) position. The stability control is turned to its maximum CCW position. The vertical-amplifier input terminal is grounded, and no horizontal-input signal is applied.

It is often helpful to supplement DC-voltage measurements with resistance measurements using a hi-lo ohmmeter. Note that although normal junction transistors "look like" open circuits to a low-voltage ohmmeter, this is not entirely true of field-effect transistors. For example,

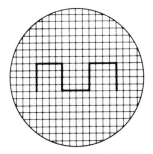

Fig. 8-8. One cycle of the square wave occupies 8 centimeters.

in Fig. 8-7, the gate terminal of Q302 normally "looks open" with re-spect to the source and drain terminals. On the other hand, the low-voltage ohmmeter will measure a finite resistance between the source and drain terminals. This resistance will measure the same value, regardless of the polarity of the applied test voltage. In the case that neither DC-voltage measurements nor resistance measurements suffice to pinpoint a defective component, substitution tests must be resorted to.

8.7 FREQUENCY-RESPONSE CHECK

A frequency-response check of a vertical amplifier can be easily made with the test setup depicted in Fig. 8-9. The video-frequency sweep gen-erator produces a pattern on the CRT screen as exemplified. Note that absorption markers are utilized to mark various frequency points along the pattern. When the video-frequency sweep signal is not demodulated, it is difficult to show frequency points with beat markers. On the other hand, absorption markers produce dips along the edges of the pattern, and are plainly visible. Some video-frequency sweep generators have built-in absorption-marker facilities. Other generators must be used with external absorption-marker boxes. If the pattern must be marked with a signal generator, the marker point can be seen in the pattern by careful inspection. It appears as a vertical line that fluctuates slightly in intensity, due to random phase variations between the sweep and marker signals.

Inadequate frequency response in vertical amplifiers can be caused by incorrectly compensated step attenuators. With reference to Fig. 8-2, a high-frequency peaking capacitor such as C106 might be incorrectly ad-justed. Again, an emitter bypass capacitor such as C103 or C105 could be open-circuited. An open drain bypass capacitor such as C4 will also impair the high-frequency response. If a coupling capacitor such as C3 becomes open, only very low frequencies can be passed from the step

207

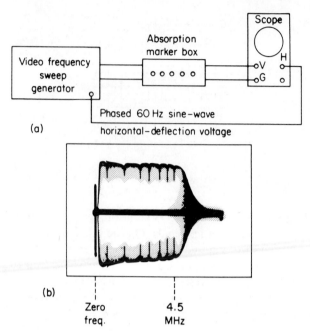

(a)

(b)

Zero freq.

4.5 MHz

Fig. 8-9. Frequency-response check of a vertical amplifier. (a) Test setup. (b) Typical video-frequency sweep display.

attenuator to Q1. Note that the optimum frequency response is uniform (flat) as approximately displayed in Fig. 8-9. If a vertical amplifier is overcompensated at high frequencies and has a rising high-frequency response, transient response will be impaired. In other words, a square wave will then be reproduced with excessive overshoot and ringing.

8.8 LINEARITY CHECK

Oscilloscope amplifiers should be precisely linear. Figure 8-10 shows a convenient method of checking the vertical and horizontal amplifiers simultaneously for linearity. The output from an audio oscillator is applied to both the vertical-input and the horizontal-input terminals. If the oscilloscope is operated on its external horizontal-input function, a diagonal trace appears on the CRT screen. Any convenient frequency may be utilized for test, such as 1 kHz, and the gain controls of the scope should be adjusted for nearly full-screen deflection. Any curvature in the trace indicates the presence of amplitude nonlinearity in either the vertical amplifier, the horizontal amplifier, or both.

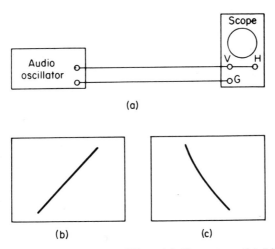

Fig. 8-10. Linearity check of scope amplifiers. (a) Test setup. (b) Linear response. (c) Nonlinear response.

To distinguish between vertical nonlinearity and horizontal nonlinearity, a 60-Hz sine-wave source may be connected to a precision decade resistance box, so that precise increments of voltage are available. The output from the decade is then applied to the vertical-input terminals of the oscilloscope. Sawtooth horizontal deflection is employed for conventional pattern display. Then, as the vertical-input voltage is increased in small steps, the pattern height is carefully observed to see whether it is increasing in direct proportion to the input voltage values. If a strict proportion is observed, the vertical amplifier is operating linearly, and the nonlinearity is in the horizontal amplifier.

Amplitude nonlinearity is generally caused by incorrect bias voltage on transistors that operate at a comparatively high level. Thus, the bias voltage for the output transistors is particularly critical. Other causes of amplitude nonlinearity include collector-junction leakage in a transistor, subnormal power-supply voltage, poor power-supply regulation, or an incorrect replacement type of transistor. Note that a certain amount of nonlinearity is often present in service-type oscilloscopes of the economy type, and cannot be corrected without extensive reworking of the amplifier circuits. On the other hand, lab-type scopes have inherently good linearity, comparable to that of high-fidelity amplifiers. However, a lab-type scope may be rated for extreme linearity only within a certain range, such as ⅔ of full-screen deflection. In any case, the specification sheet for the instrument will state the linearity limits.

INDEX